Tsunami

Engineering Perspective for Mitigation, Protection and Modelling

ADVANCED SERIES ON OCEAN ENGINEERING

Editor-in-Chief
Philip L-F Liu (*Cornell University and National University of Singapore*)

*For the complete list of titles in this series, go to https://www.worldscientific.com/series/asoe

Advanced Series on Ocean Engineering — Volume 50

Tsunami
Engineering Perspective for Mitigation, Protection and Modelling

Vallam Sundar
S. A. Sannasiraj
K. Murali
V. Sriram

IIT Madras, India

World Scientific

NEW JERSEY · LONDON · SINGAPORE · BEIJING · SHANGHAI · HONG KONG · TAIPEI · CHENNAI · TOKYO

Published by

World Scientific Publishing Co. Pte. Ltd.

5 Toh Tuck Link, Singapore 596224

USA office: 27 Warren Street, Suite 401-402, Hackensack, NJ 07601

UK office: 57 Shelton Street, Covent Garden, London WC2H 9HE

British Library Cataloguing-in-Publication Data
A catalogue record for this book is available from the British Library.

Advanced Series on Ocean Engineering — Vol. 50
TSUNAMI
Engineering Perspective for Mitigation, Protection and Modelling

ISBN 978-981-121-605-3 (hardcover)
ISBN 978-981-121-606-0 (ebook for institutions)
ISBN 978-981-121-607-7 (ebook for individuals)

For any available supplementary material, please visit
https://www.worldscientific.com/worldscibooks/10.1142/11708#t=suppl

Preface

Coastal hazards adversely affect essential infrastructures, foreshore developments and public safety. Natural disasters such as cyclones, storm surges and tsunamis can be collectively classified as coastal hazards by definition of the threat and damage they pose to the coastal regions. Tsunami is incomparable with the other perennial disasters in terms of the extent of impact experienced by the coast. Although tsunamis are rare events with a high return period, their impact cannot be totally ignored. About 80% of the tsunamis have occurred within the Pacific Ocean's "Ring of Fire", a geologically active area, where tectonic shifts make volcanic eruptions and earthquakes common.

The devastating impact of the Indian Ocean tsunami on 26 December 2004, followed by the Tohoku tsunami in 2011 and the more recent tsunami events in Indonesia, i.e., the Palu tsunami in September and the Anak Krakatau/Sunda Strait tsunami in December 2018 on the shores of the southeastern countries, recalls the importance of knowledge of coastal hazards among planners and decision makers, thereby enabling them in decision-making with regard to hazards during reconstruction and urban planning projects.

This book provides the basic physics and the understanding of tsunamis through field signature studies, physical and numerical modelling. It presents an overview of the generation and propagation of tsunami waves and their interaction with the coastal environment. The devastating effects of the tsunami in 2004 along the coastline of India in particular with the proposed and mitigation measures implemented are discussed in this book. It covers the protection against perennial coastal erosion, also taking into consideration its effectiveness in reducing the inundation run-up height and distance during the possible ingress of a tsunami.

Tsunami mitigation measures are often quite expensive which poses a challenge to mankind, since it occurs instantaneously without any prior warning. It is advised to be prepared rather than to look for solutions after its occurrence. Modelling of tsunami, i.e., both physical and numerical modelling, is an important aspect in this book as the laboratory and numerical simulations provide essential information before adopting protection measures from tsunami waves. We attempt to fulfil these requirements with the field example of the 2004 Indian Ocean tsunami along the coast of India. It also highlights the water-borne debris in the tsunami flow field, signature studies of the tsunami and the behaviour of the groyne fields.

The chapters are organized under the broad categories of overview, field study, physical modelling, numerical modelling and future consideration.

- The book is authored by experienced professors, who have undertaken extensive research in Coastal Engineering and have offered solutions to a variety of coastal engineering problems, in particular along the Indian coastline. We have carried out signature studies after the tsunami and also have prepared the master plan for tsunami mitigation measures for the two worst affected maritime states of India, i.e., Tamil Nadu and Kerala, which remain the basic documents adopted for planning.
- This book would aid students from other disciplines to understand the tsunami and its effects through signature studies.
- Both physical and numerical modelling techniques for tsunami are discussed to enhance the studies of innovative structures designed for tsunami protection and hence this book will serve as a handbook to planners, decision makers as well as beginners.

Proposed Chapters and a Brief Summary of the Chapters

The details of the chapters are discussed and presented under the following four categories, *viz.*, overview, field studies, physical modelling and numerical modelling.

The book begins with Chapter 1, Tsunami: Generation, Propagation and Effects, which provides a detailed explanation on the causes of a tsunami, its characteristics and behavioural nature. It details the transformation of a tsunami in the nearshore through diffraction, refraction, reflection and shoaling. A history of the worst tsunamis around the globe and along the Indian coast is explained with the causes and impacts. A distinct brief on the Indian Ocean tsunami (2004) is provided at the end of the chapter.

The water-borne floating debris, which is initiated and propagated along the inundation length, is a crucial parameter to be understood for the design of tsunami-resistant structures, which is discussed in Chapter 2 on water-borne debris in tsunami flow field. It presents the impact of water-borne debris on the nearshore structures during extreme events like a tsunami in three different parts. The first part contains a detailed literature survey as well as the code provisions about the debris impact on the nearshore structures due to extreme events like a tsunami. In the second part, the details of the impact tests with debris of differential weights on the scaled model residential buildings exposed to the action of different types of waves such as solitary waves, elongated single pulse waves, symmetrical and unsymmetrical N waves are discussed. In the last part of this chapter, experiments using dam-break model for generating the hydraulic bore for exact resemblance of a tsunami and its impact with the same debris and structure are presented and discussed.

Chapter 3 "Tsunami Hazards and Aspects on Design Loads" discusses the tsunami hazards, in particular the impacts of a tsunami on the coastal zone including the ports and major infrastructures such as nuclear power plants. The tsunami hazards are classified and the induced forces on coastal structures are discussed in detail with its effects and design solutions. The design loads such are hydrostatic, buoyant, hydrodynamic wave and impact forces are discussed in detail.

The stretch of the coast north of the port of Chennai situated along the southeast coast of India, an area which experienced the attack of the mighty tsunami of 2004, has been experiencing continuous erosion since the late 1960s. A number of temporary measures and a seawall as a permanent solution to counter this problem have failed. This is an area of active development. In order to combat further erosion, a detailed study of this area was conducted to suggest a suitable permanent coastal protection measure. The proposed coastal protection scheme consists of shore-connected groynes. The construction of the groyne field started in May 2004. Chapter 4 presents the behaviour of shoreline in between the groyne field based on the real-time field observations after the Indian Ocean Tsunami 2004.

Post-tsunami field surveys of the 2004 Indian Ocean tsunami have been conducted along the coasts of the two maritime states of the mainland Indian peninsula, Tamil Nadu along the east coast and Kerala on the southwest. The surveys were also carried out along the coasts of the Andaman and Nicobar Islands, where the devastating effect of the tsunami was experienced. Chapter 5 "Signature Studies (Tamil Nadu, Kerala, Andaman and Nicobar Islands)" concentrates on the propagation of tsunamis in terms of its time of arrival, run-up height and the inundation level which depend on its geographic location, bathymetry and the tsunami characteristics.

The tsunami of 2004, exhibited its might along the coast of India particularly along the Andaman and Nicobar Islands and along the two states of Kerala and Tamil Nadu in the south. Apart from the loss of valuable lives, the coastal infrastructures and existing coastal protection measures were brought down by the tsunami. This left the decision makers in a difficult position as the tsunami is a rare event and how one should respond is unclear. After the rehabilitation on a war footing, how should the planning for the future to be taken up? This led to the preparation of master plans for coastal protection for the maritime states of Tamil Nadu and Kerala affected by tsunami 2004 by us with the main idea being to take care of immediate coastal erosion problems and also to consider the effects of the tsunami wherever it was deemed essential. Although several of the measures have been implemented, only a few examples are presented and discussed in Chapter 6 "Post Facto Evaluation along Tsunami-Affected Stretches".

The use of data buoys for predicting the wave climate of sea and swell states is well known and is widely carried out in the ocean basins. A data buoy typically predicts the directional wave climate. However, the unique wave form could not be obtained from the spectral form of the data storage. In Chapter 7 "Tsunami Detection", the methodology to predict unique wave forms in particular long waves such as Cnoidal and solitary waves is explained using the time series generated in the laboratory. This has been carried out by examining the response characteristics. The details of the identification techniques from the buoy response records are comprehensively presented in this chapter.

In Chapter 8 "Effectiveness of Coastal Vegetation on Impacts due to Tsunami", the evident impact on the structures in the coastal environment from the field surveys after the great Indian Ocean tsunami of 2004 is detailed. The structures that were directly exposed to the tsunami suffered maximum damage compared to the ones fronted by vegetation on the seafront. Thus, a comprehensive experimental programme was carried out in modelling the vegetation and to investigate its interaction with the long waves representing the characteristics of a tsunami to a larger extent. The tests focussed on the measurement of dynamic pressures exerted on a wall due to Cnoidal waves in the presence and absence of a group of elastic vertical cylinders representing vegetal width. The stiffness of the vegetation appears to be one of the major parameters governing the net drag offered to the flow which required careful modelling based on its characteristics. Chapter 8 briefs the understanding of how vegetation plays an important role in terms of tsunami resistance. It presents the physical modelling aspects particularly on the vegetation and its exposure to tsunami-type flow. Based on the studies, suitable recommendations are mentioned at the end of the chapter. The extensive studies resulted in quantifying friction coefficients for tsunami and the role of vegetation in reducing inundation of coastal areas, reduction in forces on coastal infrastructure and run-up on the beaches. Detailed design methodologies are provided for coastal afforestation.

There have been lots of numerical studies on tsunamis, where special wave shapes, such as solitary waves (breaking and non-breaking), elongated solitary waves, sine waves and N-waves, have been studied. The large interest in particular on solitary waves might have two major reasons: (i) these waves have a sound mathematical background (solitary wave theory) and (ii) these waves can be quite easily generated in a flume. Chapter 9 presents a detailed discussion on the state-of-the-art experimental facilities for "real" "Tsunami Generation in Laboratory".

In order to introduce modelling aspects and nuances, the numerical modelling aspects in general are provided. Later on, the theoretical framework and numerical aspects are combined to elaborate "Tsunami Propagation Modelling" in Chapter 10. As a result, solutions of System of Shallow Water Equations are discussed under various conditions and assumptions. A computationally efficient Unstructured Explicit Finite Element Method (UEFEM) to simulate long waves in the ocean in real time over a spherical coordinate system is proposed. Later in the chapter, the various scale and computation issues are discussed in relation to requirements of different applications ranging from global scale to regional scale to synoptic scale. To demonstrate its application, the domain of the Bay of Bengal has been considered for simulation with an initial disturbance which resembles the type and location of the 2004 Indian Ocean tsunami. The wave elevation and deformations as well as time of travel of tsunami are reproduced and validated through signature studies. Numerous types of software are available around the globe for numerically modelling the propagation and behaviour of tsunamis, which are compared and discussed.

Chapter 11 "Tsunami Evolution and Run-Up" discusses the modelling aspects of the tsunami evolution and run-up characteristics using fully non-linear potential flow theory, viscous models based on particle method as well as its weakly and hybrid coupling. Based on the past events, a tsunami approaching the shore may broadly be classified as (a) a series of split waves, (b) non-breaking waves that act as a rapidly rising tide and (c) a large, turbulent wall-like wave. The modelling aspects for each of these categories using the three different modelling approaches are discussed in this chapter. Dynamics of strongly nonlinear waves including breaking, formation of shock waves and modelling the run-up aspects are discussed. Further, the chapter discusses the limitation of the one-parameter solitary wave for such a complex multi-parameter phenomenon like a tsunami using the numerical models.

Tsunamis have been often observed to approach the shore like bores depending on the bottom bathymetry. In Chapter 12 "Tsunami Impact Modelling", the tsunami front is idealized as a bore front, and the bore-induced impact pressure is evaluated in terms of the pressure impulse. The governing equation and the free surface and wall boundary conditions are cast in the finite element system to solve for the pressure impulse. The pressure impulse imposes the requirement of sudden changes in the front velocity before and after the impact on the structure. The turbulent bores, when the relative height of the bore front is large with reference to the shallow water depth in front of it, can produce large impact pressures on the structures. The influence of the bottom clearance of the elevated building on its impact is discussed. The effect of jet-type flow in comparison with bore impact has been discussed.

The contents of each of the chapters are summarized at the end.

Acknowledgements

The authors would like to acknowledge the unremitting support and unswerving encouragement from family and friends for the successful completion of this book. This book would not have been a reality without the contributions from former research scholars Dr. R. Balaji, Dr. L. Noarayanan, Dr. M.R. Behera and Dr. K.V. Anand, to name a few, from the Department of Ocean Engineering, IIT Madras, India. The field and case studies discussed in this book were the outcome of the collaborative work done with significant assistance from the Public Works Department, Tamil Nadu, and the Kerala Irrigation Department. The authors thank the research students Ms. R. Sukanya and Mr. Daniel Raj for their untiring help in drafting this book.

Last but not least, "We beg forgiveness of all those who have been with us over the course of the years and whose names we have failed to mention."

Contents

Part 2   Field Studies

Part 3 Physical Modelling

Part 4   Numerical Modelling

Chapter 10: Tsunami Propagation Modelling   215

Chapter 11: Tsunami Evolution and Run-up   229

Chapter 12: Tsunami Impact Modelling   257

PART 1

Overview

Chapter 1

Tsunami: Generation, Propagation and Effects

1.1 Introduction

A tsunami is a gravity wave phenomenon generated by submarine land-slides and earthquakes. The tsunami has a period of the order of minutes or hours. As such, even as it propagates in very deep waters it could be treated as a wave in shallow waters and hence, its phase velocity can be approximated as $(gd)^{1/2}$ and its energy is carried by this velocity. Herein, g is the acceleration constant and d is the water depth. Due to this high velocity, the energy when it reaches the coast causes disastrous effects on the coastal community. It is hence desirable to understand the mechanism of the tsunami-generating earthquakes. The largest uncertainty in tsunami modelling is derived from estimating the earthquake source parameters (Ng *et al.*, 1990; Whitmore, 1993). In the past, researchers have investigated the tsunami response along the coast for inter-plate earthquakes of different sizes and at different locations (Hebenstreit and Murty, 1989; Ng *et al.*, 1990; and Satake, 1994a). Geist and Yoshioka (1996) examined which of the parameters are the most important for controlling the generation of local tsunamis in the Cascadia subduction zone off the west coast of the U.S.

A tsunami is also characterized as a series of waves caused by the displacement of a large volume of water in the ocean. Surface waves in the ocean under normal conditions can feel the seabed only when their length is 50% of the water depth d over which they propagate, in which case the speed with which they travel will not be a function of the water depth and will be a function of only the period. The main difference between an ocean wave and a tsunami lies in the magnitude of the length or period. When a long wave or a tsunami propagates towards the shore, its length and speed reduce.

For an ocean wave, its period T will be up to about 30s and its length L will be of a few hundred metres. It feels the seabed when d/L is less than 0.5 due to which it undergoes deformation. The main cause for the generation of waves is the wind acting over the ocean surface. A tsunami in simple terms is said to behave as a shallow water wave in very deep waters of say a few kilometres and propagates with its length of up to a few kilometres with its period being a few hours and not in seconds, as in case of normal waves. Among the series of waves in the case of a tsunami as stated earlier, the first wave need not necessarily be the shortest. If the shoreline recedes drastically, it signifies the trough of a tsunami wave approaching the shore initially, and this phenomenon is called the drawdown; this results in a temporary sudden drop of the sea level. The drawdown is always immediately succeeded by the crest of tsunami. The sea-level upsurge in the vertically upward direction during its propagation is known as the run-up. When a tsunami wave is generated, it usually ends up causing huge damage to the coastal front and on many occasions results in loss of life and destruction of property. The damage in particular at locations of flatter beach slope is expected to be devastating. When the flooding tsunami waves retreat they tend to carry loose objects and people out into the sea. Though its impact is restricted to coastal area, its effects can be so severe that the entire ocean basin is disturbed and deviates from normality. Prediction of the occurrence of a tsunami is impossible and only a postmortem of its effects in terms of loss of life and property can be done. As prevention is not possible, it is only the preparedness that is extremely important to minimize the damage. A signature study is usually conducted during the post-tsunami period to assess the inundation height and distance as these would serve as useful data for validating numerical models that are capable of simulating a tsunami of different characteristics. This would possibly give an idea on its destruction capability. A "tsunami" is sometimes incorrectly referred to as a tidal wave, which in fact is generated due to a large increase in the gravitational attraction between the celestial bodies, *viz.*, sun, moon and the earth. The storm surge is a natural coastal hazard that occurs due to the formation of low pressure and converging winds resulting in a cyclone associated with sea-level rise and a surge resulting in large-scale destruction along the coast. Sufficient warning prior to the occurrence of a storm surge helps in minimizing the damage, whereas this is impossible in the case of a tsunami as stated earlier. Tsunamis are neither connected with the weather nor with the tides. "*Tsunami*" is a Japanese word which translates as "harbor wave" and sometimes as "seismic wave".

1.2   Causes of a Tsunami

A tsunami is likely to be generated either by underwater earthquakes that account for about 85% of the tsunamis around the globe (Whitmore *et al.*,

2008), or huge landslides or volcanic eruptions that lead to a displacement of a huge mass of water in the ocean. In general, a tsunami occurs due to any seismic changes in the Earth's crust near or in the seabed. Tsunamis are generated by four distinct mechanisms:

- The most common type being the abrupt deformation of the seabed due to tectonic earthquakes along subduction boundaries. A pictorial representation of a tsunami due to displacement of tectonic plates causing earthquake is shown in Fig. 1.1(a).
- Submarine landslides either triggered by earthquakes or local instability of the continental slope.
- Major volcanic eruptions which displace a sufficient mass of water.
- Underwater manmade explosions with subsequent water inrush.

The main causes for the generation of a tsunami are illustrated in Fig. 1.1(b). The impact of foreign bodies such as meteors causes water displacement. The intensity of these is not as devastating as the seismic factor. A large volume of water is displaced by falling debris and can lead to a tsunami. The energy contained in a tsunami wave dissipates rapidly as it approaches the coast in the form of high impact force with very large wave heights and greater extent of inundation into the coastal terrain.

1.3   Tsunami Earthquakes

1.3.1   *General*

Tsunami earthquakes were originally identified as having an anomalous tsunami excitation corresponding to an excitation of short-period (1–20 s) body and surface seismic waves. The cause of the discrepancy is thought to be the characteristic long source duration which provides low excitation of short-period seismic energy (Kanamori, 1972, 1977). For this reason, moment magnitude (Mw) is thought to be a more diagnostic magnitude scale for tsunami earthquakes, mb is the body-wave magnitude scale and the tsunami earthquake discrepancy has recently been cast as a discrepancy between Ms (intensity of body surface waves) and M~ (Kanamori and Kikuchi, 1993). Other factors that relate to anomalous tsunami excitation as described by Pelayo and Wiens (1992) include very shallow dip of the source mechanism and low rigidity at the source. These characteristics relate to the fact that most tsunami earthquakes are located within the accretionary wedge seaward of the normal seismic front of a subduction zone. Fukao (1979) proposed that faults within the accretionary wedge may

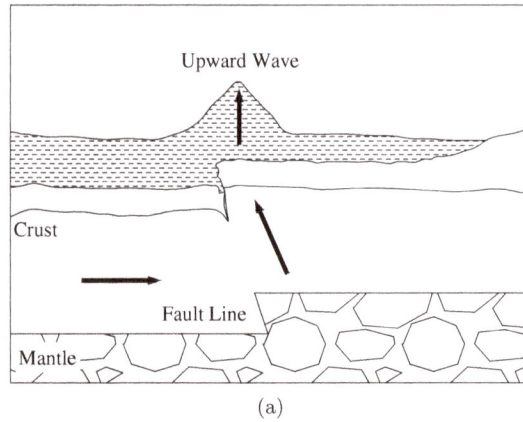

(a)

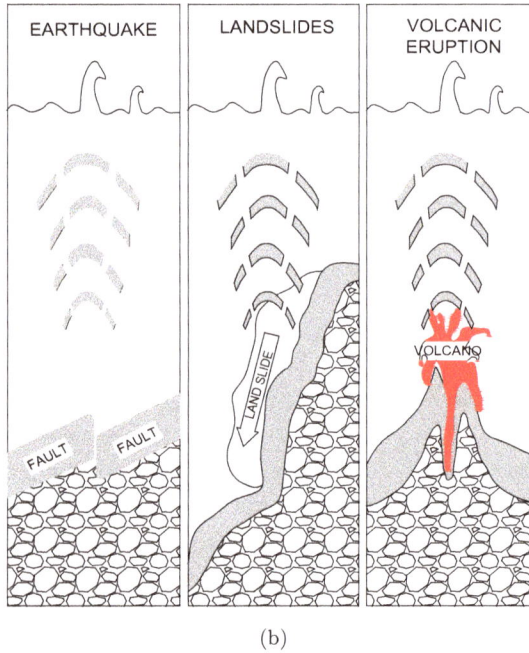

(b)

Fig. 1.1   (a) Pictorial representation of tsunami due to earthquake. (b) Pictorial representation of causes for a tsunami.

be stressed towards failure after a large inter-plate earthquake as exemplified by the 1963 Kurile tsunami earthquake. However, tsunami earthquakes do not always follow large inter-plate events. Alternatively, Pelayo and Wiens (1990) suggested that slow tsunami earthquakes may occur in the subduction zone segments that lack large earthquakes and segments with a low ratio of seismic to aseismic slip. Fukao (1979) has also suggested a popular model for the kinematics of tsunami earthquakes, which documents a focal mechanism change during rupture that is thought to represent an upward branching from the inter-plate thrust. Shallow coseismic rupture greatly enhances the vertical displacement at the seafloor. Moreover, because the shear modulus of sediments in the accretionary wedge is typically low, tsunami earthquakes are probably characterized by higher amounts of slip, for a given scalar moment, than earthquakes within more consolidated rock. This is borne out by the study of the 1992 Nicaragua tsunami earthquake by Satake (1994a), where for the same seismic moment, the tsunami source model calls for six times the amount of slip in low-rigidity rocks compared to the conventional seismic source model.

It is useful to tabulate typical values for source depth, amount of slip and source duration for selected tsunami earthquakes around the world (Table 1.1), which shows the discrepancy between short-period magnitude determination (mb and Ms) and Mw, the shallow source depth, and the long source duration of the events. In addition, these earthquakes are associated with large amounts of slip relative to normal subduction zone earthquakes of a similar magnitude (~0.5–2.5 m, Lay *et al.*, 1982). It is hypothesized that great subduction zone earthquakes (Mw ~9) are likely to cause large tsunamis.

Measurement of mb has changed several times. As originally defined by Gutenberg and Richter (1945) mb was based on the maximum amplitude of waves in the first 10 s or more. However, the length of the period influences the magnitude obtained. Early USGS/NEIC practice was to measure mb on the first second (just the first few P-waves), but since 1978 they measure the first 20 s. The modern practice is to measure short-period mb scale at less than 3 s.

1.3.2 *Aspects of earthquake rupture that affect tsunamis*

It is difficult to define a direct relationship between the parameters influencing earthquake rupture and the local tsunami event. The magnitude of earthquake which resulted in a tsunami is a good measure to determine the size of the tsunami that has travelled far from the origin of the earthquake. However, additional details are required to calculate the magnitude and run-up of the local tsunami. A variety of parameters are used to define the nature of earthquake ruptures that can vary across space and time.

Table 1.1.   Source parameters for notable tsunami earthquakes.

Location	Date	mb	Ms	Mw	Depth (km)	Rupture duration (s)	Slip(m)	Reference
Peru	20 November 1960	NA	6.8	7.6	9	130	2.3-3.4	Pelayo and Wiens (1990)
Kurile	20 October 1963	NA	7.2	7.8	9	85	3.3-5.0	Pelayo and Wiens (1992)
Kurile	10 June 1975	5.8	6.8	7.5	5	60	3.8-5.7	Pelayo and Wiens (1992)
Nicaragua	2 September 1992	5.3	7.2	7.6-7.7	0-10	100	—	Satake (1994b)
Flores	12 December 1992	6.7	7.5	7.8	3-15	60	3.2-9.6	Imamura and Kikuchi (1994)
Sumatra	26 December 2004	—	—	9.3	30	500-600	~15	Sundar et al. (2007)

1.3.2.1 *Amount of average slip*

A fault displacement either in the vertical and/or horizontal direction with respect to another is witnessed in the event of a tsunami generation. The average slip is the distance of displacement of two adjacent sides averaged over the rupture area. For an earthquake rupture, the relation between fault slip and permanent seafloor offset is linear. In other words, if the average slip of an earthquake event (E1) is over twice in magnitude of another such individual/independent earthquake event (E2), their corresponding seafloor offset and initial tsunami would also differ by a factor of two. In view of the numerous nonlinear effects governing tsunami wave propagation, the relation between the average slip and amplitude of the tsunami near shore is not exactly linear.

The magnitude of the slip throughout the rupture area of the earthquake has the highest influence on the size of the local tsunami. Generally, the slip increases with the magnitude of the earthquake, although sans the knowledge about rupture area, physical properties of rocks and other allied parameters, it is impossible to accurately arrive at the magnitude of slip. The relationship between magnitude of earthquake and average slip that resulted in a tsunami or otherwise as per Geist and Yoshioka (1996) is projected in Fig. 1.2, from which it is understood that the average slip during rupture of a tsunami earthquake seems to be larger than a non-tsunami earthquake of the same magnitude.

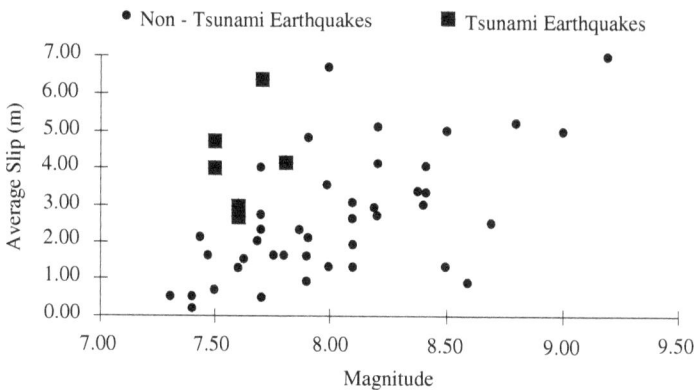

Source: Geist and Yoshioka (1996).

Fig. 1.2   Relationship between average slip and magnitude of earthquakes.

1.3.2.2   *Depth of rupture*

The depth of the earthquake rupture is an important criterion in determining the size of the local tsunami event. Shallow rupture on the Earth's surface results in a large degree of seafloor offset and thereby a larger initial tsunami compared to one due to deep rupture earthquake. An example is shown in Fig. 1.3. The left-hand side of Fig. 1.3 shows the portion of a fault that ruptures as indicated by dotted lines. The local tsunami that is generated from this rupture is shown as a synthetic marigram (wave amplitude as a function of time). Fault C, shown in Fig. 1.3(b), ruptures much shallower in the Earth and generates a substantially larger tsunami.

1.3.2.3   *Orientation of slip vector*

Thrust faulting is illustrated in Fig. 1.4, where the overlying block moves upward and over the underlying block. Analysing the same three

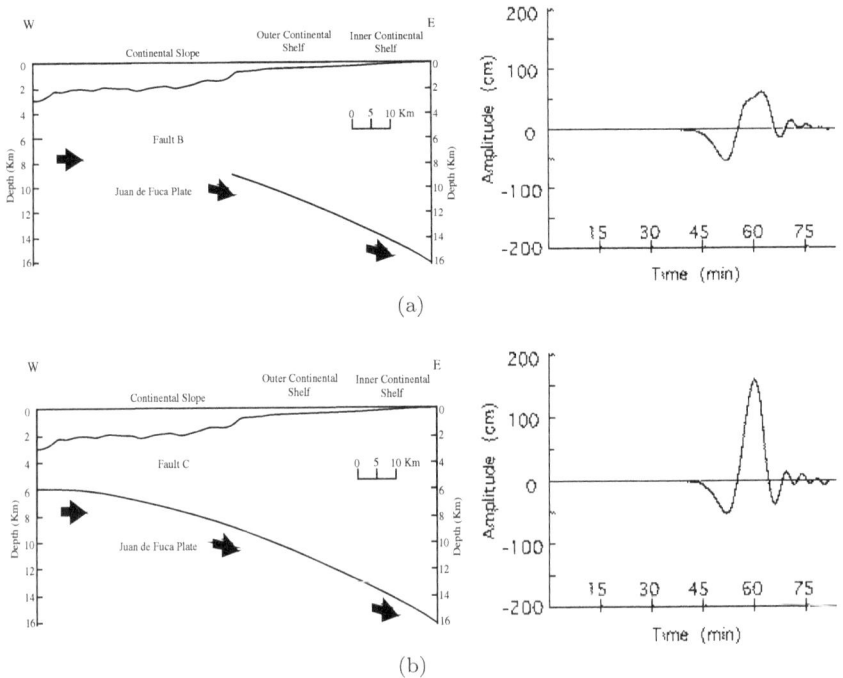

Source: Geist and Yoshioka (1996).

Fig. 1.3   Initial water surface displacement for different rupture types. (a) Test case 1: Rupture on Fault B. (b) Test Case 2: Rupture on Fault C.

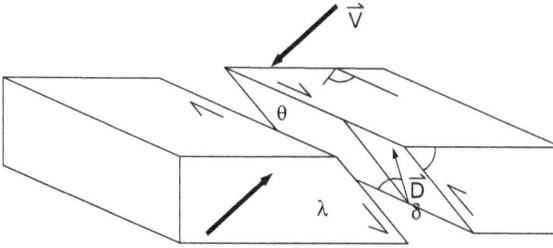

Fig. 1.4   Definition of slip vector orientation.

dimensions, the fault block could move in and out of the page direction as well (as indicated by the arrows). Oblique faulting is more prone to occur in a subduction zone when the fault/plate moving downward slides at an oblique angle (θ) relative to the overriding plate/fault. The obliquity of the slip vector (D) in the fault plane of dip (δ) is measured by the angle (λ) that the slip vector makes with a horizontal line in the fault plane.

The details of the rupture are important in the generation of a local tsunami event. When a rupture arises following the oblique faulting as described earlier, the vertical offset of the seafloor is substantially different from a simple thrust faulting. This results in the generation of secondary tsunamis that initially travel in a different direction than the deep-ocean and local tsunamis.

1.3.2.4   *Duration of rupture*

A tsunami is a gravity wave and its period is an important property that determines its generation and propagation characteristics out of the source region. This period should be directly related to the rupture duration. It is to be noted that the initial disturbance of the sea surface, if very fast, may not produce a tsunami as this will be related to short-period waves and such a wave cannot carry a lot of energy. In this case, the earthquake may not produce a tsunami. On the contrary, if the rupture duration is long enough, the transfer of energy to the water column may be so slow that a tsunami or a wave is not generated.

The rupture duration that is capable of generating tsunamis is of the order of 60–600 s, based on past experience (Table 1.1). In this case, there seems to be a sufficiently long time available for the kinetic energy of the slip to be transferred to the water column and conversion of the same into potential energy by means of initial disturbance. Once this energy conversion has taken place, the initial disturbance of the water level at the source region forms into a gravity wave and propagates out of the source region.

1.4   Tsunami due to Volcano

Volcanic tsunamis can result from violent submarine explosions. They can also be caused by *caldera* (in Spanish it means a basin-shaped volcanic depression, at least 1.6 km in diameter) collapses, tectonic movement from volcanic activity and flank failure in a water source. The wave initiated at deep waters moves in a vertical direction and attains great speeds of about 650 mph, whereas in shallow waters it could still retain a speed of 200 mph. The energy/power carried along with the waves does not reduce when they approach the land, and thus extreme wave heights and longer distance of inundation are observed along the coast. Tanguy *et al.* (1998) pointed out that ∼5% of tsunamis are formed from volcanoes and ∼16.9% of volcanic fatalities occur from tsunamis.

Although relatively infrequent, violent volcanic eruptions also represent impulsive disturbances, which can displace a great volume of water and generate extremely destructive tsunami waves in the immediate source area. According to this mechanism, waves may be generated by the sudden displacement of water caused by a volcanic explosion, due to a volcano's slope failure, or more likely by a phreatomagmatic explosion and collapse/engulfment of the volcanic magmatic chambers. One of the largest and most destructive tsunamis ever recorded was generated on 26 August 1883 after the explosion and collapse of the volcano of Krakatoa (Krakatau) in Indonesia. This explosion generated waves that reached 40 m, destroyed coastal towns and villages along the Sunda Strait in both the islands of Java and Sumatra, killing 36,417 people. It is also believed that the destruction of the Minoan civilization in Greece in 1490 B.C. was caused by the explosion/collapse of the volcano of Santorin in the Aegean Sea. The major tsunami volcano events in the recent past are given in Table 1.2.

1.5   Appearance of a Tsunami Wave

When a tsunami approaches the land, its appearance and behaviour depend on the topography of the seafloor, the actual shape of the shoreline, reefs, bays, river entrances and the slope of the beach. Occasionally, a tsunami may tend to break in the offshore region or may form into a bore, which resembles a wave step with steep breaking front. Bore formation can result when the tsunami wave propagates from deep waters into shallow bay or

Table 1.2.   Major tsunamis due to volcanic eruption.

Year	Location	Impact and cause
1979	Illiwerung, Indonesia	☐ Volcanic landslide ☐ 9 m waves
1980	St Helens, Washington, USA	☐ Volcanic landslide ☐ 250 m waves in Spirit Lake
1983	Illiwerung, Indonesia	☐ Submarine eruption
1986	Nyos, Cameroon	☐ Underwater CO_2 eruption ☐ 75 m tsunami waves
1988	Vulcano, Italy	☐ Volcanic landslide ☐ 5.5 m waves
1994	Rabaul, Papua New Guinea	☐ Pyroclastic flow induced ☐ 1.2 m waves
1996	Karymsky, Russia	☐ Phreatomagmatic eruption ☐ 30 m waves
1997	Soufriere Hills, Montserrat	☐ Volcanic debris slide ☐ 3 m waves
2002	Stromboli, Italy	☐ Landslide-induced tsunami
2007	Ritter, Papua New Guinea	☐ Eruption-induced landslide

Source: Wikipedia.

river. The wave heights can be accentuated when concentrated over headlands or while entering enclosed bay areas with a wide entrance that narrows down progressively. Coral reefs can act as buffers to dissipate the energy and decrease its impact on the shoreline. A wind swell may ride over a tsunami wave thereby increasing its height furthermore.

The degree of damage causing wave activity differs vastly along a coast. The inundation of a tsunami wave may go up to even a couple of kilometres or more covering a wide area of land mass with water and debris. Flooding resulting from tsunami tends to carry floating objects and people when they retreat. The maximum vertical run-up of a wave as it hits the shore may rise over 10 m depending upon local geomorphological features.

1.6   Tsunami Characteristics

The wave length (L) of the tsunami is decided by the displaced water mass during its origination. A large displacement of water mass (proportional to the displaced underwater land mass) generates a tsunami wave since the energy propagates as an oscillating wave in a medium such as water. The wave length is set longer relative to wave amplitude since the horizontal distribution of energy (from the source such as fault plane) is larger than the vertical transfer of momentum against gravity. Having stated this, the

restoring force for a tsunami wave is gravity. As per the long wave theory
for a gravity wave, the speed of the wave (C) can be deduced.

$$C = \frac{L}{T} = \sqrt{gd}. \tag{1.1}$$

$$L = T\sqrt{gd}, \tag{1.2}$$

where T is the tsunami wave period, g is the acceleration due to gravity
and d is the water depth. Following the aforementioned relation, one can
understand that in given water depth, if the wave length is longer, the wave
period should also be higher. The two parameters (L and T) have strong
interdependency in a constant water depth. In general, the period is of the
order of 10 min to few hours.

The speed, wavelength of the tsunami in different water depths from
offshore along with that for a longest expected ocean wave with $T = 30\,\text{s}$
projected in Table 1.3 provides a clear picture of the characteristics of a
tsunami.

It is seen that while a tsunami propagates in water more than 1 km deep,
the average speed of the wave is more than 350 km/h. The wave reaches
the continental shelf, say in the case of Bay of Bengal with the origin along
off the coast of Sumatra, in about a few hours. However, the wave speed

Table 1.3.   Comparison between a tsunami and gravity wave.

Depth d (m)	Tsunami ($T = 2\,\text{h}$)		Probable longest ocean wave ($T = 30\,\text{s}$)		
	Wave length, L (km)	Wave celerity, C (km/h)	Wave length, L (km)	Wave celerity, C (km/h)	
6,000	436.75	873.5	1.404	168.62	
5,000	398.65	797.3	1.404	168.62	
4,000	356.565	713.13	1.404	168.62	Deep
3,000	308.79	617.58	1.404	168.62	Water
2,000	252.15	504.3	1.404	168.62	
1,000	178.3	356.6	1.404	168.62	
750	154.39	308.29	1.404	168.62	
750	149.16	298.32	1.397	167.73	
500	126.065	252.13	1.375	163.38	Intermediate
250	89.14	178.28	1.210	134.76	Water
100	56.375	112.75	0.87	82.033	
50	39.865	79.73	0.635	79.73	Shallow
10	17.825	35.65	0.295	35.65	Water

substantially reduces in the continental shelf of the order of 100 km/h and takes an average 2 h to make a landfall along the east coast of India.

The tsunami wave speed in a very shallow depth is further accelerated by its elevation (η) following the relation:

$$C = \sqrt{g(d + \eta)}. \tag{1.3}$$

The aforementioned relation is used to estimate wave speed if the tsunami elevation is comparable to the water depth.

Another important aspect of tsunami propagation is its propagation over longer distances without any loss of energy. A long gravity wave propagates without loss of energy over longer distances which makes a tsunami wave easily sweep the regional seas.

1.7 Tsunami Transformation

1.7.1 *Shoaling*

While, the waves propagate from deep to shallower region, the wave length decreases as stated earlier. The conservation of energy leads to an increase in wave amplitude. The increase in wave amplitude is called shoaling. In general, the tsunami is unnoticeable in deep waters, i.e., the amplitude is of the order of few centimetre to less than a metre. Due to the decrease of wave length of the order of 10 folds, the wave amplitude increases more than 1 m up to 10–20 m. The tsunami amplitude (η_2) in a relatively shallower location (d_2) can be calculated while a tsunami amplitude (η_1) in deeper water d_1 is known.

$$\eta_2 = \eta_1 \left(\frac{d_1}{d_2}\right)^{1/4}. \tag{1.4}$$

From the aforesaid equation, it can be inferred that the tsunami wave amplitude of 0.5 m while at the entrance of the continental shelf with a water depth of 200 m amplifies to 1 m in a water depth of 10 m. In most of the context, the tsunami height is represented at 5–10 m depth for most of the design calculations. In a water depth of 5 m, the wave amplifies to about two-and-half times. Figure 1.5 presents the wave evolution from an origin (water depth), say 0.5 m new born wave in a water depth of 2000 m towards the coast. It is assumed here that no loss of energy occurs during the wave propagation. However, if the wave height could not be sustained, it breaks and the wave approaches the shore as a bore. This effect is considered later to estimate the wave impact force.

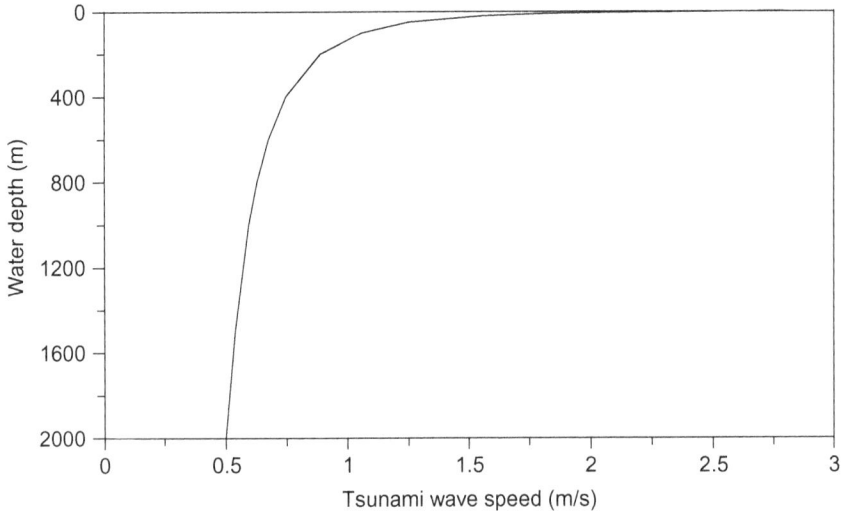

Fig. 1.5   Long wave evolution from deep water to near shore.

1.7.2   *Diffraction, refraction and reflection*

Similar to a wind wave, the tsunami wave is subjected to phenomena such as diffraction, refraction and reflection. The wave diffracts while it encounters an island or headland-like projection, e.g., the southernmost Indian coast acts as a headland for the tsunami encounter. The variation in bathymetry refracts the wave, in which the change in wave direction happens according to Snell's law. According to the conservation of energy principle, as the direction of wave changes it results in reduced width between consecutive waves and the wave heights between consecutive waves increases. Similar increase in wave heights also results after shoaling when the water depth decreases near the shore. A steeper coast further reflects substantial part of wave energy compared to a flatter slope, in which the wave spends its energy for its propagation and deformation.

By considering the aforesaid wave phenomena, the southernmost tip of the Indian subcontinent is a precarious region since the tsunami energy gets concentrated due to refraction. However, the presence of the Sri Lankan island diffracted the tsunami leading to the net energy concentration from the southern tip of Kanyakumari to about 50–60 km on the west coast of India. This is the combined effect of diffraction and refraction. The energy concentration due to such a combined effect is shown in Fig. 1.6.

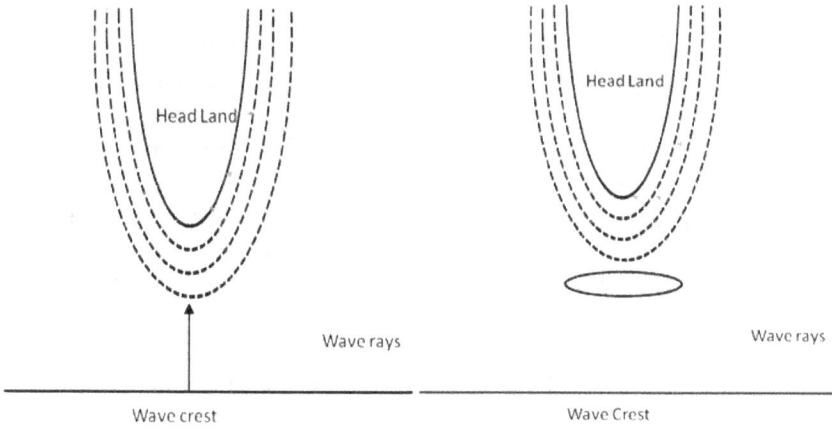

Fig. 1.6 Concentration of wave energy due to refraction and combined diffraction and refraction.

The tsunami energy is also focussed in a converging ocean basin such as say V-shaped.

1.7.3 Nearshore tsunami deformation

While tsunami approaches the nearshore region, it changes its profile by the way of transforming its energy. The potential head transforms to kinetic energy, in which the wave either breaks or moves like a broken bore or undular bore. Figure 1.7 depicts the abovementioned deformation process. The factors affecting the deformation process are water depth, beach slope, wave height and length modulated by the bottom bathymetry. A mild slope beach produces an undular bore, while a relatively shorter tsunami approaches the coast. In this way, the tsunami does not lose its energy while making a landfall and possibly causing a huge impact. The bore-like formation can induce impulsive impact forces on the port and coastal structures and might cause large destruction of the structure. The bore formation might also occur at the tsunami landfall and the propagation speed can be approximately estimated by the following:

$$C = \beta\sqrt{gd}, \tag{1.5}$$

where β is the ground roughness parameter and takes the value of 0.7 for a very rough surface and 2.0 for a very smooth surface. To make a conservative design, the earlier value of wave speed can be taken as particle

(a)

(b)

(c)

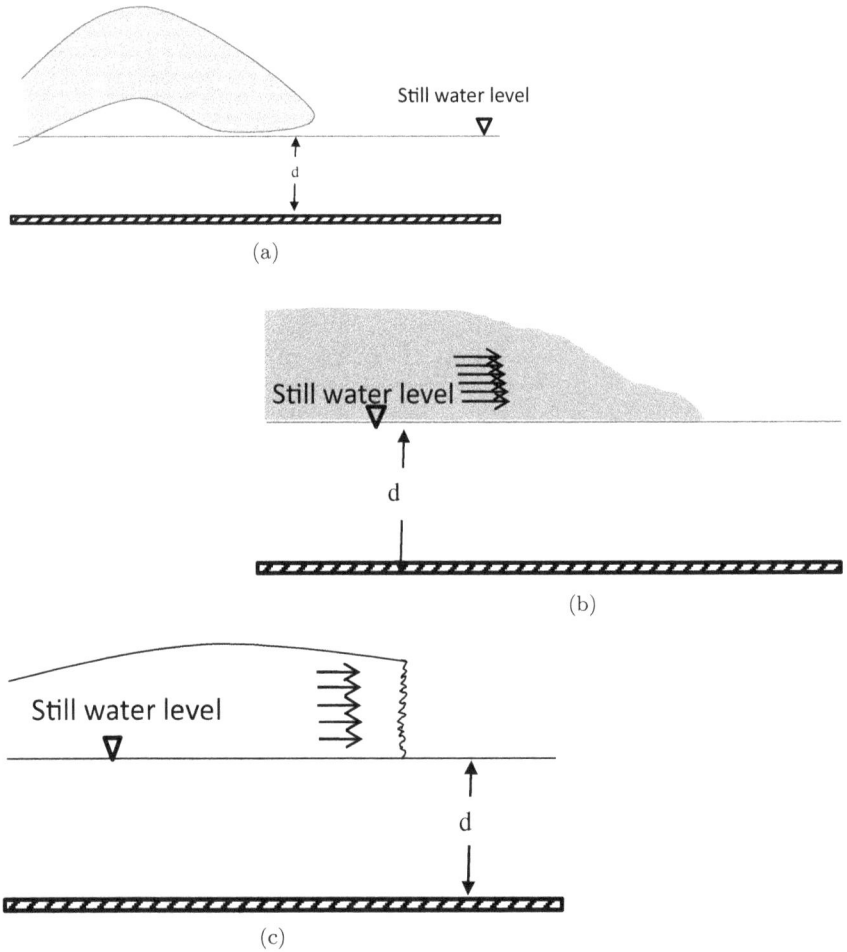

Fig. 1.7   Tsunami wave deformation near the coast. (a) Breaking wave front. (b) Undular bore. (c) Broken bore.

velocity. However, this would lead to an uneconomical design, if the tsunami is very infrequent.

1.7.4   *Tsunami wave dispersion*

Unlike ocean surface waves, tsunami waves are non-dispersive. It can be clearly understood from Eq. (1.5) that the speed of the wave is independent of its period, i.e., either a short or a long wave, the tsunami speed is a function of only water depth. However, the tsunami wave disperses similar to solitary wave dispersion in very shallow waters. It is essentially due to the development of wave nonlinearity expressed in terms of H/d ratio. In this aspect, the wave period also plays an important role. The wave speed in this case can be derived from the second-order nonlinear wave theory.

$$C = \sqrt{gd}\left\{1 - \frac{(kd)^2}{6}\right\}, \tag{1.6}$$

where k is the wave number ($k = 2\pi/L$). The wave dispersion happens in relatively shallow water and in particular when a tsunami propagates from deeper water to the continental shelf. This leads to the splitting of tsunami waves into multiple waves and hence the coast is exposed to multiple landfalls even from a single generation source.

1.8   Occurrence of Recent Devastating Tsunamis

1.8.1   *General*

The Pacific basin region is more susceptible to submarine faulting and subsequent tsunami events owing to the numerous volcanic activities and tectonic plate slips. On the other hand, the Indian and Atlantic Ocean belts are far less geologically active, with a few exceptions.

Ninety tsunamis were recorded in the Pacific since 1900 (every decade saw one major devastating tsunami). A brief description of the few most important tsunamis that have occurred around the globe is provided in Table 1.4.

On 11 March 2011, an earthquake of magnitude 9.1 shook the Pacific coast off Tohoku coast, Japan. The epicentre of the earthquake was ∼70 km east of the Oshika Peninsula at Tohoku and its hypocentre ∼29 km below the sea level. The 2011 Tohoku earthquake was the most powerful earthquake recorded in Japan, and the fourth most powerful earthquake in the world since 1900. The earthquake triggered powerful tsunami waves that reached heights of up to 40.5 m at Miyako in Tōhoku's Iwate Prefecture, and which, in the Sendai area, travelled up to 10 km inland. Apart from resulting in excessive damage to life and property, it also resulted in nuclear accidents at the Fukushima Daiichi Nuclear Power Plant complex. Residents within a 20 km radius of the Fukushima Daiichi Nuclear Power Plant and a 10 km radius of the Fukushima Daini Nuclear Power Plant were evacuated.

The most recent tsunami that occurred in 2018 is often referred to as Palu tsunami, the brief details of which are presented herein. On

Table 1.4.   Most disastrous tsunamis around the globe.

Location	Date	Earthquake magnitude (Richter scale)	Tsunami impact
Pandeglang, Java Island, Indonesia	22 December 2018	NA, caused due to volcanic eruption	Run-up heights — 3 m
Palu, Indonesia	29 September 2018	7.5	Wave run-up up to 6.1 m was observed near the coast
Tohoku, Japan	11 March 2011	9.0–9.1	Run-up heights — 39 m Inundation distance — 10 km
Sumatra, Indonesia	26 December 2004	9.1	Run-up heights — 50 m Inundation distance — 5 km
North Pacific Coast, Japan	11 March 2011	9.0	Run-up heights — 10 m Wave speed — 800 kmph
Lisbon, Portugal	1 November 1755	8.5	Run-up heights — 30 m
Krakatau, Indonesia	27 August 1883	NA, caused due to volcanic eruption	Run-up heights — 37 m
Enshunada Sea, Japan	20 September 1498	8.3	Tsunami waves along the coasts of Kii, Mikawa, Surugu, Izu and Sagami. The waves were powerful enough to breach a spit, which had previously separated Lake Hamana from the sea
Nankaido, Japan	28 October 1707	8.4	Run-up heights — 27 m
Sanriku, Japan	15 June 1896	7.6	Run-up heights — 38.2 m
Northern Chile	13 August 1868	8.5	Run-up heights — 21 m, lasted between two and three days
Ryuku Islands	24 April 1771	7.4	Run-up heights — 11–15 m

28 September 2018, an earthquake with a magnitude of 7.5 on the Richter scale shook the Sulawesi island, just after 5:02 pm Western Indonesian Time (WIT), resulting in the formation of tsunami waves which hit the Palu bay area. Apart from Palu, Donggala, Parigi Moutong and Sigi were few of the most affected coastal stretches of the island. Balaroa and Petobo were the two villages which were not inundated by the tsunami, but massive sand liquefaction owing to a high impact earthquake destroyed the area by sinking the entire infrastructure.

The initial observation was made at 2 pm WIT with an earthquake of magnitude 6.1 followed by a series of temblors. The location experienced 27 such aftershocks before it succumbed to an intense 7.5 magnitude earthquake about 6 miles/9.6 km deep (*Source*: U.S. Geological Survey). The Indonesian natural disaster agency, Badan Nasional Penanggulangan Bencana (BNPB), reported that the power outages due to earthquakes crippled the communication of the possible forthcoming disaster to the masses. The progress direction of the earthquake from its origination point is projected in Fig. 1.8.

The volcanic eruptions from Anak Krakatau resulted in the deadly tsunami waves which propagated through the Sunda Strait between Java and Sumatra islands on the night of 22 December 2018. The active volcano was emitting smoke and lava since June 2018, which eventually lead to a major landslide (an area of 64 hectare/158 acre) on its southwest side. This in turn triggered a tsunami that propagated predominantly towards the strait, the details of which were reported by the European Space Agency's

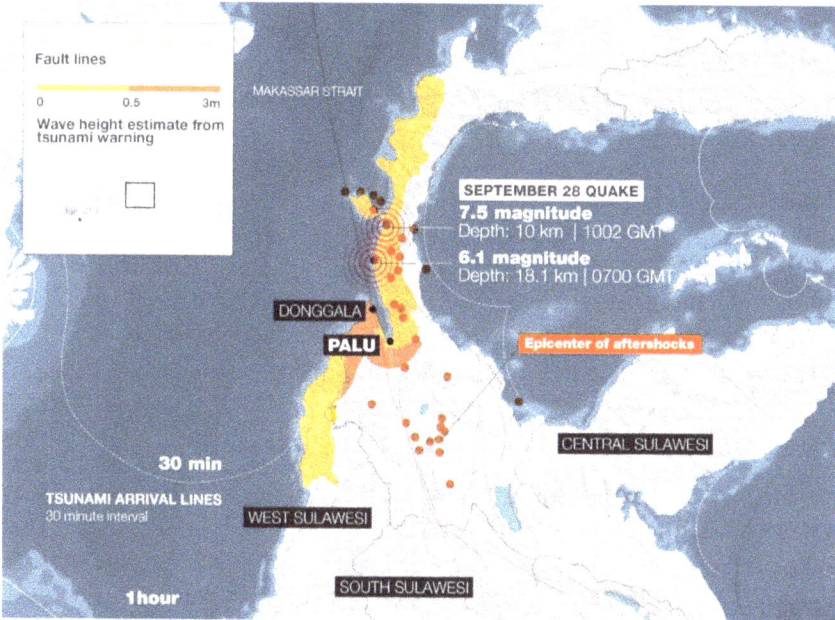

Source: Reuters; USGC; Indonesia Tsunami Early Warning System; BMKG.

Fig. 1.8   The direction of propagation of the earthquake of Palu tsunami from its origination point.

Fig. 1.9   The location of Anak Krakatau from the Indonesian islands.

Sentinel-1 satellite. The aforesaid volcanic eruption occurred at about 60 km off the coast, as shown in Fig. 1.9, and hence it propagated quickly without a warning resulting in a huge loss of life and property. The Pandeglang region in the Java Island is estimated to be the most affected stretch.

Experts and volcanologists suggest that recurrences of such tsunami events are likely to take place in the vicinity of Anak Krakatau till the volcano remains active. Since its proximal location to the densely populated Indonesian islands, such generated tsunami waves will travel faster and hit the coast within a short notice, not providing sufficient time for evacuation.

1.8.2 *Tsunamis in India*

A tsunami was recorded along the Sri Lankan coastline in the year 1883. It is believed that this was a result of a volcanic eruption at Krakatoa leading to about 1 m high surge along the coastline. Table 1.5 shows the details of tsunami occurrences in India. An account on the past tsunamis along the Indian coast has been discussed in detail by Rastogi and Jaiswal (2006).

In general, there are two sources of tsunami origination which could possibly affect the Indian coast. One is at Mekran coast that has in the past affected the northern west coast, particularly the Gulf of Cambay

Table 1.5.   Tsunamis recorded in India.

Date	Cause	Impact
326 B.C.	Unknown	Unknown. Noted from the history Alexander the Great
April–May 1008	Earthquake along Iranian coast	Unknown
31 December 1881	A 7.9 Richter scale earthquake beneath Car Nicobar	Entire east coast of India and Andaman and Nicobar Islands; 1 m tsunami was recorded at Chennai.
27 August 1883	Explosion of the Krakatoa Volcano in Indonesia	East coast of India was affected; 1.5 m tsunami at Chennai; 0.6 m at Nagapattinam; 0.2 m at Arden.
26 June 1941	An 8.1 Richter scale earthquake in the Andaman archipelago at 12.9°N, 92.5°E	East coast of India was affected with 0.75–1.25 m tsunami. Some damage was reported.
27 November 1945	An 8.5 Richter scale earthquake at a distance of about 100 km south of Karachi	West coast of India from north to Karwar was affected; 12 m tsunami was felt at Kandla.
26 December 2004	A 9.1 Richter scale earthquake at Sumatra coast	East coast of India, Andaman and Nicobar Islands, Lakshadweep islands, Kerala coast were recorded 1–7 m tsunami. 9,700 people lost their lives and 6,000 people were reported to be missing. Minor damage to coastal and harbour structures.

along the Gujarat coast. Gulf of Khambhat is also prone to be exposed to concentration of energy towards inner gulf. The other origin is along the fault line in between the Indian and Burmese plates along which the great Indian Ocean tsunami originated.

1.8.3   *Indian Ocean tsunami of 26 December 2004, in brief*

1.8.3.1   *Origin*

The attack of tsunami on the Indian coastal waters was a great surprise, even though tsunamis are not entirely unknown in India. On 26 December 2004, a massive earthquake of magnitude 9.3 on the Richter scale occurred at 3.316°N, 95.854°E off the Sumatra coast of Indonesia at 7:58:53 am local

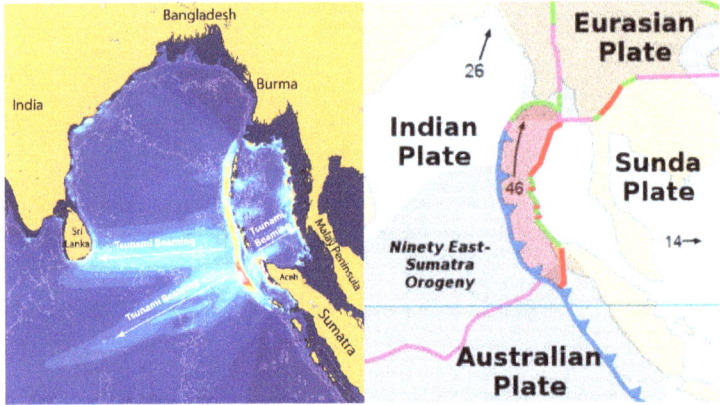

Fig. 1.10   Fault line of Indian Ocean tsunami.

time. This unusually large earthquake occurred due to slipping of 1,200 km of fault line by about 15 m along the subduction zone, where the India Plate subducted under the Burma microplate. The interface between the two plates results in a large fault, termed an inter-plate thrust or mega thrust. Figure 1.10 shows the fault line in the tectonic plates. Given the large extent of the slip, it did not happen instantaneously but took place in two phases over several minutes. Seismographic and acoustic data suggest that the earthquake happened in two phases. The first phase involved formation of a rupture of about 400 km long and 100 km wide, located 30 km beneath the sea bed (http://en.wikipedia.org/). Beginning off the coast of Aceh, Indonesia, the rupture proceeded at a speed of about 2.8 km/s or 10,000 km/h for duration of about 100 s. In the second phase, after a pause of about 100 s, the rupture continued northwards towards the Andaman and Nicobar Islands. However, the northern rupture proceeded only at about 2.1 km/s for another 5 min until meeting a plate boundary where the fault changes from subduction to strike-slip. This reduced the speed of the water displacement thereby reducing the magnitude of the tsunami that hit the northern part of the Indian Ocean (Charles *et al.*, 2005).

As the fault line of length 1,200 km was oriented along the north–south direction, the tsunami wave effectively propagated along the east–west direction. Thus, Bangladesh situated in the northern end of the Bay of Bengal, suffered from limited causalities despite being low lying. It was spared because the tremor waves propagated at a much slower rate along the northern rupture zone, thereby reducing its energy.

1.8.3.2 *Propagation*

Given the large extent of the slip, it did not happen instantaneously but took place in two phases over several minutes. Seismographic and acoustic data suggest that the earthquake happened in two phases. The first phase involved formation of a rupture of about 400 km long and 100 km wide, located 30 km beneath the seabed. Beginning off the coast of Aceh, Indonesia, the rupture proceeded at a speed of about 2.8 km/s or 10,000 km/h for duration of about 100 s. In the second phase, after a pause of about 100 s, the rupture continued northwards towards the Andaman and Nicobar Islands. However, the northern rupture proceeded only at a rate of about 2.1 km/s for about 5 min until it coincided with the plate boundary, where the fault line propagates from subduction to strike-slip. This effect reduced the rate of water displacement, they by diminishing the magnitude of tsunami, which hit the northern coast of Indian Ocean.

In addition to the horizontal movement between the plates, the seabed is also said to be vertically deformed by several metres — thereby triggering the devastating tsunami. The waves rather emanating from a point source got transmitted over the entire 1,200 km length of the rupture leading to a greater extent over which the waves were noted, reaching as far as Mexico, Chile and the Arctic. It is expected that many islands in that regions would have displaced both horizontally and vertically. Through the geophysical changes observed via satellite imagery and GPS measurements, it was found that the Andaman and Nicobar Islands have shifted southwest by about 4 m. The aforesaid techniques do not yield accurate predictions in the case of vertical shift. However, the tidal stations give a clue and this will be discussed in following sections.

The Indian Ocean tsunami with a speed of about 800 km/h at its source of generation initially struck off the western coast of northern Sumatra, Indonesia, had exhibited its might along the coasts of India, Sri Lanka, Malaysia, Indonesia, Thailand, Maldives and even up to Bangladesh and Somalia. The Indian Ocean tsunami reached a height of up to about 10 m along the coast and is believed to be the worst coastal disaster leading to the loss of several thousands of lives, property and disrupting the normal life for a few months. Along the Indian coast, the maritime states of Tamil Nadu, Kerala, Puducherry and Andhra Pradesh were the worst affected by the tsunami that claimed 18,000 lives. The Survey of India has noted that the tsunami hit Cuddalore at 8.00 am (IST), Chennai at 8.40 am (IST), and Paradip, Machilipatnam and Vishakhapatnam at 9.00 am (IST). The time taken by the tsunami to travel from east coast of Sri Lanka to reach the Indian shore was just 10 min.

1.8.4   *Initial warning*

There is no pre-warning of the occurrence of tsunami unlike in the case of the generation a storm surge. The initial warning could be the receding of the sea several metres into the ocean in which case, several metres of the stretch of the coast will be exposed. This in fact during the great Indian Ocean tsunami of 2004 has driven people with curiosity towards ocean not knowing that their lives will be lost within a few seconds. In shallower beach slopes, the extent of recession can exceed beyond a few hundreds of metres. People along the Pacific are aware of this phenomena as the occurrence of a tsunami is not that rare an event compared to the Indian Ocean region where people were unaware of the danger at the shore and the warning signs of a propagation of a tsunami. Out of the seven shore temples constructed around 4–9th century in Mahabalipuram, (12.6269°N, 80.1927°E) situated in the southeast coast of India, also identified by UNESCO as a world heritage site, it is believed that six of them have been sacrificed to the ocean due to erosion. During the receding of water into the ocean to an extent of 500 m during the tsunami of 2004, some of the hidden relics were exposed as can be seen in Fig. 1.11.

1.9   Tsunami Warning System

The first tsunami warning system was established for the Hawaiian Islands in 1948. The primary objective was through the well-monitored seismographs to detect as well as to locate the existence of all possible earthquakes that can cause a tsunami. The warning system was also to assess the effect of tsunami along the coast through the measurement of water level variations through tide-gauging stations along the Pacific; and to evaluate the time of arrival of the tsunami to serve as a warning to follow well-defined evacuation procedures. The first Deep-ocean Assessment and Reporting of Tsunami (DART) station was installed in 2001 with six tsunami detection buoys along the northern Pacific Ocean coast by NOAA (Meinig *et al.*, 2005). The DART station constitutes pressure sensors mounted on the seafloor referred to as bottom pressure recorder (BPR) and a surface buoy. The BPR is proficient in identifying a tsunami wave with an amplitude as low as 1 cm in a water depth of 6000 m (Eble and González, 1991). On a continuous basis, the pressure sensors will be transmitting the pressure due to the

Fig. 1.11   Ancient structures in Mahabalipuram exposed due to the receding of water along the coast during tsunami 2004.

propagating surface ocean waves to the buoy. These signals from the buoy are transmitted through an acoustic link to a satellite, which in turn sends the information to the ground station from where, as and when a tsunami is detected in the deep ocean, dissemination of the warning becomes easier. As a tsunami is very long period wave, the pressure exerted by which is many times greater, it is easy to differentiate between a normal wave and a tsunami which forms the basic parameter for identifying a tsunami generated in the deep ocean. The schematic representation of the DART system is shown in Fig. 1.12.

The Indian Tsunami Early Warning System (ITEWS) at the Indian National Centre for Ocean Information Services (INCOIS), Hyderabad, established in October 2007 monitors real-time earthquake activity throughout the Indian Ocean to evaluate potential tsunami-genic earthquakes. The ITEWS constitutes a real-time network of seismic stations, BPRs, 30 tide gauge stations to detect tsunami-genic earthquakes, to monitor tsunamis and to provide timely tsunami warning due to any major earthquake in the Indian Ocean in less than 10 min to Andaman and Nicobar

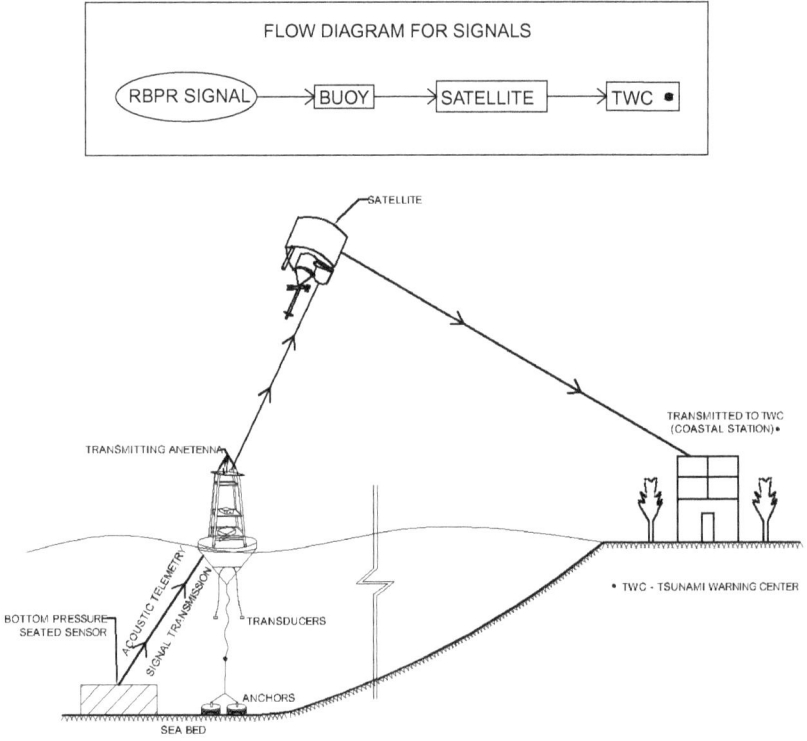

Fig. 1.12   Schematic representation of the DART system.

Islands and a few hours to the mainland. For details refer to Nayak and Srinivasa Kumar (2011).

1.10 Summary

In geographical scale, the tsunami impact is felt only along the shoreline and may inundate regions up to 3 km landwards. Nevertheless, the magnitude of its destructive forces is greater when compared to other type of disasters. Efficient planning with sound economic investments should be made on (implementing/deploying) a (warning/evacuation) system that can influence the safety of the coastal stretches against such disasters.

References

Charles, J. A., Chen, J., Hong-Kie, T., David, R., Sidao, N., Vala, H., Hiroo, K., Thorne, L., Shamita, D., Don, H., Gene, I., Jascha, P., and David, W.

(2005). "Rupture process of the 2004 Sumatra–Andaman earthquake." *Science*, 308(5725), 1133–1139.

Devi, E. U., Sunanda, M. V., Ajay Kumar, B., Patanjali Kumar, Ch., and Srinivasa Kumar, T. (2016). "Real-time earthquake monitoring at the Indian Tsunami Early Warning System for tsunami advisories in the Indian Ocean." *The International Journal of Ocean and Climate Systems*, January–April, 20–26.

Eble, M. C., and González, F. I. (1991). "Deep-ocean bottom pressure measurements in the Northeast Pacific." *Journal of Atmospheric and Oceanic Technology*, 8(2), 221–233.

Eisner, R. K. (2001). "Designing for tsunamis: Seven principles for planning and designing for tsunami hazards." National Tsunami Hazard Mitigation Program, p. 60.

Fukao, Y. (1979). "Tsunami earthquakes and sub-duction processes near deep-sea trenches." *Journal of Geophysical Research*, 84, 2303–2314.

Geist, E., and Yoshioka, S. (1996). "Source parameters controlling the generation and propagation of potential local tsunamis along the Cascadia margin." *Natural Hazards*, 13, 151–177.

Gutenberg, B., and Richter, C. F. (1946). "Earthquake study in southern California." *Eos, Transactions American Geophysical Union*, 28(4), 633–634.

Hebenstreit, G. T., and Murty, T. S. (1989). "Tsunami amplitudes from local earthquakes in the Pacific Northwest region of North America." *Part 1: The Outer Coast, Marine Geodesy*, 13, 101–146.

Imamura, F., and Kikuchi, M. (1994). "Moment Release of the 1992 Flores Island Earthquake Inferred from Tsunami and Teleseismic Data." *Sci. Tsunami Hazards* 12, 67–76.

Kanamori, H. (1972). "Mechanism of tsunami earthquakes." *Physics of the Earth and Planetary Interiors*, 6, 346–359.

Kanamori, H. (1977). "Seismic and aseismic slip along sub-duction zones and their tectonic implications" in M. Talwani and W. C. Pitman III (eds.), *Island Arcs, Deep Sea Trenches and Back-Arc Basins*, Vol. 1. American Geophysical Union, Washington, D.C., 163–174.

Kanamori, H., and Kikuchi, M. (1993). "The 1992 Nicaragua earthquake: a slow tsunami earthquake associated with subducted sediments." *Nature* 361, 714–716.

Lay, T., Kanamori, H., and Ruff, L. (1982). "The asperity model and the nature of large sub-duction zone earthquakes." *Earthquake Prediction Research*, 1, 3–71.

Meinig, C., Stalin, S. E., Nakamura, A. I., González, F., and Milburn, H. G. (2005). "Technology developments in real-time tsunami measuring, monitoring and forecasting." In Oceans 2005 MTS/IEEE, 19–23 September 2005, Washington, D.C.

Nayak, S., and Srinivasa Kumar, T. (2011). "Tsunami watch and warning centers" in Harsh K. Gupta (ed.), *Encyclopedia of Solid Earth Geophysics*, Vol. 2. Springer, Dordrecht, 1498–1505.

Ng, M. K., LeBlond, P. H., and Murty, T. S. (1990). "Simulation of tsunamis from great earthquakes on the Cascadia sub-duction zone." *Science*, 250, 1248–1251.

Pelayo, A. M., and Wiens, D. A. (1990). "The November 20, 1960 Peru tsunami earthquake: Source mechanism of a slow event." *Geophysical Research Letters*, 17, 661–664.

Pelayo, A. M., and Wiens, D. A. (1992). "Tsunami earthquakes: Slow thrust-faulting events in the accretionary wedge." *Journal of Geophysical Research*, 97, 15321–15337.

Rastogi, B. K., and Jaiswal, R. K. (2006). "A catalog of tsunamis in the Indian Ocean." *Science of Tsunami Hazards*, 25, 128–143.

Satake, K. (1994a). "Numerical computation of tsunamis from a hypothetical Cascadia Earthquake (abs.)." *Seismological Research Letters*, 65, 25.

Sundar, V., Sannasiraj, S. A., Murali, K., and Sundaravadivelu, R. (2007). "Run-up and inundation along the Indian peninsula, including the Andaman Islands, due to Great Indian Ocean Tsunami." *Journal of Waterway, Port, Coastal, and Ocean Engineering*, 133(6), 401–413.

Tanguy, J. C., Ribiere, C. H., Scarth, A., and Tjetjep, W. S. (1998). "Victims from volcanic eruptions: A revised database." *Bulletin of Volcanology*, 60, 137–144.

Whitmore, P., Benz, H., and Bolton, M. (2008). "NOAA/West Coast and Alaska Tsunami Warning Center Pacific Ocean response criteria." *Science of Tsunami Hazards*, 27(2), 1–21.

Whitmore, P. M. (1993). "Expected tsunami amplitudes and currents along the North American coast for Cascadia sub-duction zone earthquakes." *Natural Hazards*, 8, 59–73.

Chapter 2

Tsunami-Driven Debris and Its Impact

2.1 Introduction

Tsunami waves from the deep ocean amplify in the near-shore zone and often inundate the shore with high velocity. The local beach slope dictates the inundation distance and height. Owing to the very long wavelength associated with tsunami waves, they behave as shallow water waves in the deep ocean. During the progress of a tsunami towards the shore, all the objects lying over the seabed are lifted and carried away along with these waves. Once the wave approaches the shore, it drives forward all the objects lying in its path and carries along with it. The objects thus driven are called tsunami-driven debris. According to ASCE/SEI 7-16, virtually anything in the flow path of a tsunami that can float given the inundation depth and that cannot withstand the water flow becomes debris. Large objects which are supposed to be in the marine environment have been often found hundreds of metres inland after a devastating tsunami.

This tsunami-driven debris may have its origin either onshore or off-shore. The onshore debris driven during tsunami differs from location to location. Dwelling units, gas stations, harbours, falling trees, parked vehicles, important infrastructures, industrial units and more importantly nuclear power plant facilities along a stretch of a cost form the main source for the onshore debris. Coastal areas nearer to these commercial places are often flooded with debris like fishing boats, fishing nets, nuclear power plant disposals, cars, wooden logs, container and several other objects as can be seen in Fig. 2.1. However, offshore debris is not site specific and includes all the objects starting from plastic materials to floating buoys, damaged marine vessels, coral reefs, dead animals, etc. The kind of offshore debris carried during tsunami depends on the epicentre of the tsunami and the direction in which it propagates towards the shore.

During the inundation of tsunami waves towards the shore, they propagate as bores near the shoreline. The last part of the tsunami after breaking is called a bore. This bore carries all the offshore and onshore debris and

Fig. 2.1   Floating debris during the 2004 tsunami in Indian Ocean at Chennai, India.

collides with the near-shore structures. This causes an additional impact loading to the structures, the effect of which would not have been incorporated in the design. The debris-induced loading can be quite large and could often be one of the greatest causes for the damage of the structures in the coastal zone. This may lead to the failure of local components of the structure or failure of the whole structure. Thus, the failure of the structures also results in building waste such as concrete, steel, and timber, which form as debris, being lifted and carried along with the tsunami bore. The image of a damaged dwelling/housing unit due to tsunami which ultimately reduces to debris is shown in Fig. 2.2. The kind of building waste depends on the material used for the construction of such buildings. Thus, a wide range of debris is generated during a tsunami which makes modelling of the debris very difficult.

2.2   Outline of Damage Caused by Debris

The degree of damage to a building depends on the type of debris (flexible or rigid), type of material used for construction, type of construction, distance from the sea and orientation of the building to the tsunami. Wall-type structures are bound, resulting in more volume of debris. In the case of pile-supported structures, the tsunami waves are expected to pass through

Fig. 2.2   Damaged structures due to impact loading from tsunami.

them without much damage. The impact force from the debris is felt locally at any given point/surface area of the building, where the dimensions of the debris are smaller compared to that of the building. Field observation of the tsunami-affected areas has shown that the impact of debris on structures plays a vital role in their failure (Yeh *et al.*, 2014). Mooring ropes/cables may loosen or break causing the ship to drift along with these tsunami waves. Since, ships are naturally designed to float; they get easily carried away along with these tsunami waves when inundation depth is more than the draft of the ship. The motion of the ship depends on the intensity of the tsunami waves, especially wave velocity and wave height. Thus, coastal structures very near to the ports and fishing harbour have more possibility of damage due to ships and boats. Small boats may get capsized and sink, crash with berthing structures or be left stranded over the top of the berthing structure. As the height of the inundating tsunami waves increases, large ships, containers, automobiles, etc., are carried away along with the waves.

2.3   Different Types of Debris

The debris generated during tsunamis can be broadly classified as aerial floating debris, sub-aerial floating debris and submerged debris, which are briefly discussed in the following sections.

2.3.1 *Aerial floating debris*

This type of debris is usually large, having unit weight greater than water. It is found afloat in the surging bore progressing landward from the sea and its motion is hindered by the seabed slope as can be seen in Fig. 2.3. Thus, this kind of debris usually has velocity lesser than the bore velocity. Debris like automobiles, containers, boats and ships with design draught higher than the inundation depth of the bore usually exhibits the aforementioned characteristics.

2.3.2 *Sub-aerial floating debris*

This kind of debris usually floats in water. The debris is initially accelerated by the incoming tsunami bore and after travelling some distance the debris gains velocity and moves along with the surging bore since no hindrance is imposed to its motion. This kind of debris usually has enough keel distance during floating such that the boundary effect from the bottom does not affect its motion as shown in Fig. 2.4. Debris like wooden logs and boats usually exhibits the aforementioned characteristics.

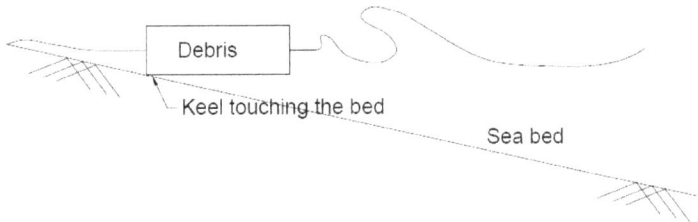

Fig. 2.3   Aerial floating debris.

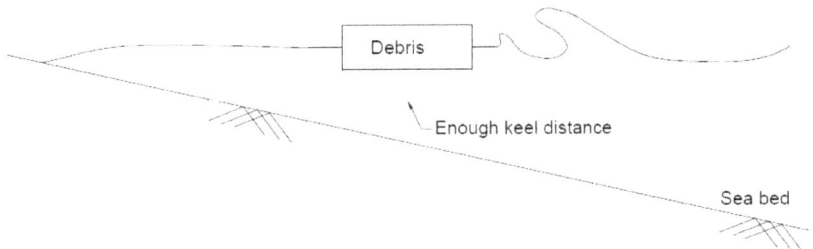

Fig. 2.4   Sub-aerial floating debris.

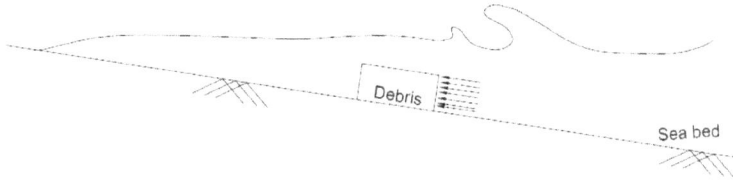

Fig. 2.5   Submerged debris.

2.3.3   *Submerged debris*

This kind of debris does not float in water because of its huge mass and large unit weight and they usually move along the bed, the motion of which being initiated as long as the horizontal force on it is higher than any kind of friction offered. The debris like boulders and concrete blocks usually exhibits the aforesaid characteristics which are schematically represented in Fig. 2.5.

2.4   Boulders as Debris

Research has also been carried out by considering stones and boulders as debris. In Geology, a boulder is a rock with grain size of usually no less than 300 mm or 12 inch in diameter (ASTM D2487-11, 2011). A wide range of classification of boulders is reported in the literature. The broad classification of boulders is based on their volume or mass, long or intermediate axis or even both of the aforementioned parameters. Several authors have used a variety of terms to describe boulder deposits like clasts, megaclasts, megaliths, etc. (Paris *et al.*, 2011). Hence, there is no standard set of rules to define the boulder deposits, which indirectly affects our knowledge or understanding.

Blair and McPherson (1999) gave a modified Udden–Wentworth grain-size scale and defined everything greater than 256 mm intermediate axis diameter as boulders. Paris *et al.* (2011) have summarized the boulder accumulations and classified them as per the following morphological zones:

(1) **Offshore boulders:** These boulders are found in the infra-tidal region and are frequently moved by wave action and are broken, reformed and sometimes transported onshore.
(2) **Platform boulders:** These boulders are located on inter-tidal and supra-tidal shore platforms, which consist of fields of scattered boulders,

clusters of imbricated boulders and accumulations of boulders in pools carved in the platform.

(3) **Boulder ridges:** These are the boulder deposits above the highest astronomical tide level.

(4) **Boulder beaches:** These are formed at the cliff platform junction. They differ from boulder ridges in that boulder beaches are found mostly in the inter-tidal region, whereas boulder ridges are found above the highest astronomical tide level. Boulder ridges as well as boulder beaches are found to be formed mostly by storm action.

(5) **Cliff-top boulders:** These are deposited at very high elevations. Formation of cliff-top boulder deposits requires a very high storm wave or a tsunami.

2.5   Research on Motion and Impact of Debris

After the 2004 Indian Ocean Tsunami and 2011 Japan Tsunami, debris-driven impact study has increased (Ko *et al.*, 2015; Nistor *et al.*, 2009). It is difficult to study the influence of debris impact on structures as there is a wide variation in the debris mass, debris shape, debris material and the number of debris which the structure attracts. Previous research on the impact of water driven debris on the near shore structures depicts that the impact force caused by debris is much larger compared to the force due to the hydraulic bore alone (Nouri *et al.*, 2010). This water driven debris induces a kind of impact force which is generally larger in magnitude and induces impulsive loading. The maximum force acting on a structure is influenced by debris parameters (shape, size, orientation, deformability and debris velocity, tsunami bore height and tsunami velocity and the type of structure itself whether flexible or rigid) (Haehnel and Daly, 2004). Most of the debris impact study focussed on finding the debris impact force on fixed structures such as buildings, columns in buildings and bridge piers.

Matsutomi (1999) performed small- and full-scale experiments on the impact forces generated by driftwood impacting rigid structures. The full-scale experiments correspond to the wooden logs impacted a frame conducted in open air and impact forces were measured. An empirical formula for estimating the impact force, F, was reported (Eq. 2.1).

$$\frac{F}{\gamma_w D^2 L} = 1.6 C_M \left(\frac{u}{\sqrt{gD}}\right)^{1.2} \left(\frac{\sigma_f}{\gamma_w L}\right)^{0.4}, \qquad (2.1)$$

where γ_w is the specific weight of wood, D is the diameter of the log, L is the length of the log, C_M is a coefficient which depends on the flow passing around the receiving wall (1.7 for bore or surge, and 1.9 for steady flow), u is the velocity of the log at impact and σ_f is the yield stress of the log. Later Matsutomi (2009) modified the formula for estimating the collision force with velocity of debris as flow velocity and C_M as 1.9.

$$\frac{F}{\gamma_w D^2 L} = 3.8 \left(\frac{h_r}{D}\right)^{0.6} \left(\frac{\sigma_f}{\gamma_w L}\right)^{0.4}, \tag{2.2}$$

where h_r is the inundation depth at the back of the structure estimated from the maximum-height traces at the front and back of the structure.

Haehnel and Daly (2004) carried out experiments on the woody-debris impact on the structures in the Cold Regions Research and Engineering Laboratory's Ice Engineering Facility by flood-induced hydraulic bore. A single degree of freedom was adopted for evaluating the effective stiffness of the debris and the structure which is an additional parameter needed for obtaining an analytical solution for the impact force of debris using contact stiffness approach.

$$F = u\sqrt{km}, \tag{2.3}$$

where u is the debris velocity, k is the effective stiffness between the structure and the debris, m is the mass of the debris. The stiffness and mass of the debris is determined based on the type of debris. Three approaches, *viz.*, work energy approach, contact stiffness approach and impulse-momentum approach for finding debris impact load were reviewed. Each of these approaches requires an additional parameter to be defined like the stopping distance for work energy approach, the effective stiffness during collision for contact stiffness approach and the stopping time (i.e., contact duration) for impulse-momentum approach. The interdependence of stopping distance, stopping time and the effective stiffness is discussed.

Nouri *et al.* (2009) carried out series of experiments over circular and diamond shaped structures to quantify the forces and pressures exerted on them due to the impact of a tsunami-induced hydraulic bore and extended the tests on the impact force of debris on the structure using two wooden logs and concluded that the debris after hitting the structure can bounce back and hit the structure with the magnitude of impact during the second impact being lesser.

Madurapperuma and Wijeyewickrema (2012) examined the impact of a tsunami-borne shipping container on a reinforced concrete column using finite-element analysis and reported that the peak contact forces increased almost linearly with an increase in the container velocity.

Naito *et al.* (2012) classified the debris in coastal regions into small, moderate and large debris based on four quantitative characteristics: mass, stiffness, size and buoyancy. The categorization was framed within the context of the magnitude of forces that could be generated by its impact on a structure. The basic assessment procedure for determining the type of debris that may be generated in case of tsunami and also capacity of the debris to cause potential damage to structure was also provided.

Stolle *et al.* (2015) studied the dynamics of the debris flow in a tsunami as the dynamic characteristics of the debris motion are crucial in identifying the high-risk areas. Experiments with 1, 3, 9 and 18 of smart debris for understanding the trajectories of multiple debris motion in tsunami or any flood event were carried out. The smart debris used was capable of collecting information of the debris motion in all the 6 degrees of freedom. From the experiments, it was found that multiple pieces of debris lead to debris collision which resulted in spreading of debris in the surging bore. In case of a piece of single debris, it travelled along with the surging bore with lesser spreading angle. Goseberg *et al.* (2017) proposed a novel method of identifying the motion of the debris using Radio frequency identification (RFID) technique. This study on debris dynamics could be helpful in the probabilistic assessment of debris impact on structure in the surging tsunami flow. Recently, Stolle *et al.* (2017) developed a probabilistic approach for the transport of debris. Although probabilistic models on debris transport are still in the research stage, this study helps in the basic understanding on the spreading of debris.

2.6   Guidelines for Debris Impact Loading

Prior to December 2004, the design of structures for tsunami was not given due importance, as such an occurrence has a high return period. Improving the design guidelines for structures near the coastline took prominence as a result of the devastating effect of tsunami 2004. Majority of design codes do not quantify the impact forces close to reality owing to uncertainties in defining the wave characteristics and a lack of knowledge in understanding the underlying physical processes and as such is difficult to estimate the impact of the debris on structures as it is influenced by the mass, velocity, draft and the orientation of the debris with respect to the structure.

In the earlier stages of research on the debris impact during a tsunami (i.e., during 2005) ASCE/SEI 7-05(2005) does not specify how to incorporate tsunami and tsunami-driven loading in the design of tsunami-resistant buildings. The hydrodynamics is mostly explained in perspective of flooding. It provides an analytical solution for finding the debris impact force based on impulse momentum approach with the relation given by Eqs. (2.4) and (2.5). It provides information regarding the debris in different locations and also specifies the weight of the debris to be considered for calculating its impact. The design of buildings subjected to impact loading due to debris is restricted to category III and category IV structures during the design flood, where the resulting damage is very high in terms of fatalities and infrastructure. For the design, debris of mass 450 kg having contact duration of 0.03 s is adopted for estimating and incorporating the debris load, which is felt appropriate, although it is site specific. Later ASCE/SEI 7 included tsunami load as a part of structural load in tsunami design zone. ASCE/SEI 7-16(2016) included a chapter on tsunami loads and effects that explains in detail the various aspects like tsunami wave forms, surging bore velocity. It provides different load cases based on inundation depth of the tsunami bore. As per ASCE/SEI 7-16, category III and category IV structures have to be checked also for tsunami load combinations.

$$I = \int_0^\tau F \, dt = d(mu); \quad \tau \to 0. \tag{2.4}$$

$$F = \frac{\pi W V C_I C_O C_D C_B R_{\max}}{2g\Delta t}. \tag{2.5}$$

In the aforementioned equations, I is impulse momentum, F is impact force, V is the velocity of object, W is debris weight, g is the acceleration due to gravity, Δt is the impact duration (time to reduce object velocity to zero), C_I is the importance coefficient, C_O is the orientation coefficient, C_D is the depth coefficient, C_B is the blockage coefficient, $R_{\max}$ is the maximum response ratio for impulsive load.

As per the recommendations of ASCE/SEI 7-16, design of buildings should incorporate the debris impact load, whenever, the inundation depth is more than 0.914 m (3 ft). The design instantaneous debris impact force based on contact stiffness approach provided by ASCE/SEI 7-16 is given as follows:

$$F_R = I_{\text{tsu}} C_O U \sqrt{K m_{U_{\max}}}, \tag{2.6}$$

where I_{tsu} is the importance factor for structure, C_O is the orientation coefficient, $U_{\max}$ is the maximum flow velocity at the site, k is the effective stiffness of the impacting debris or the lateral stiffness of the impacted structural element deformed by impact, whichever is less and m_d is the mass of the debris in kilograms. The obtained impact force is converted into equivalent static load by multiplying with dynamic response ratio. The static load is applied as a point load on the structure. The maximum load imposed by different debris along with its point of application on the structure is given in Table 2.1 and the duration of the impact of different types of construction materials is provided in Table 2.2.

Since shipping containers, ships and barges induces a larger impact load, the consideration of this load in the design of category III and category IV structures should be done only after detailed assessment of hazard at site.

Another such guideline for design of structures for vertical evacuation during a tsunami is provided in FEMA P-646. It adopts the contact stiffness approach for finding the debris impact force as adopted by ASCE/SEI 7. The mass, stiffness and hydrodynamic mass coefficient for

Table 2.1.   Debris type and maximum impact force.

Debris type	Maximum impact force	Point of application
Submerged tumbling boulder and concrete debris	$36 \times I_{\text{tsu}}$ (KN)	At 0.61 m from grade level
Vehicle	$130 \times I_{\text{tsu}}$ (KN)	From any point greater than 0.914 m to inundation depth
Shipping containers	$980 \times I_{\text{tsu}}$ (KN)	For a wall, the impact shall be assumed to act along the horizontal centre of the wall
Extraordinary debris impacts by ships and barges	Weight = Lightship weight + 30% of Dead weight tonnage	Applied anywhere from base of the structure up to 1.3 times the inundation depth + height of the deck of vessel

Table 2.2.   Duration of force due to impact of debris.

Type of construction	Duration (t) of Impact (s)	
	Wall	Pile
Wood	0.7–1.1	0.5–1.0
Steel	NA	0.2–0.4
Reinforced concrete	0.2–0.4	0.3–0.6
Concrete masonry	0.3–0.6	0.3–0.6

Table 2.3.   Mass and stiffness of some water-borne floating debris.

Type of debris	Mass (m_d) in kg	Hydrodynamic mass Co-eff. (c)	Debris stiffness (K_d) in N/m
Lumber or Wood log-oriented longitudinally	450	0	2.4×10^6*
20-ft standard shipping container-oriented longitudinally	2200 (empty)	0.30	85×10^6**
20-ft standard shipping container-oriented transverse to flow	2200 (empty)	1.00	80×10^6**
20-ft heavy shipping container-oriented longitudinally	2400 (empty)	0.30	93×10^6**
20-ft heavy shipping container-oriented transverse to flow	2400 (empty)	1.00	87×10^6**
40-ft standard shipping container-oriented longitudinally	3800 (empty)	0.20	60×10^6
40-ft standard shipping container-oriented transverse to flow	3800 (empty)	1.00	40×10^6

Note: *Haehnal and Daly (2002); **Peterson and Naito (2012).
Source: FEMA (2012).

common water-borne debris is provided in Table 2.3. As the possibility of two or more debris impacting simultaneously is less, it need not be considered.

During tsunami propagation, vertical evacuation structure should be preferred as its speed will be very high for resisting and hence escape route would be towards a higher plane or structure. This is all the more important in the event debris floating around which has been highlighted by FEMA P-646.

2.6.1   *Debris damming force*

As per FEMA P-646, the damming effect caused by accumulation of water-borne debris can be treated as a hydrodynamic force enhanced by the breadth of the debris dam against the front face of the structure. The debris damming force is given as follows:

$$F_{dm} = \frac{1}{2}\rho C_d B_d (hu^2)_{\max}, \qquad (2.7)$$

where ρ is the fluid density, C_d is the drag coefficient, B_d is the breadth of the debris dam, h is flow depth, and u is flow velocity at the location of the structure, C_d is drag coefficient. The maximum momentum flux per unit width $(hu^2)_{max}$ should be obtained by executing a numerical simulation model or acquiring existing simulated data. If no numerical simulation results are available, an estimate of $(hu^2)_{max}$ can be determined using Eq. (2.8).

$$(hu^2)_{max} = gR^2 \left(0.125 - 0.235\frac{z}{R} + 0.11 \left(\frac{z}{R} \right)^2 \right), \qquad (2.8)$$

where g is the acceleration due to gravity, R is the design run-up elevation and z is the ground elevation at the base of the structure.

The debris damming effect represents the accumulation of debris along the structure and the total forces exerted can be resisted by a number of structural components, which depend on the dimensions and size of the structural frame and debris dam. The debris damming force, F_{dm}, is assumed to be uniformly distributed load acting over the extent of debris dam. An appropriate tributary width is ought to be assigned to each of the structural component resisting the force and uniformly distributed over submerged height of the component.

FEMA55 (2011) provides guidelines in terms of design consideration for coastal flooded structures. It also provides formula for finding lower and upper bound values for the design flood velocity. It adopts impulse momentum approach for determining the debris impact loading. FEMA states that the impact force due to debris on structure is influenced by its location in the potential debris stream, specifically if it is:

- closer to a building or obstruction or in a location likely to be washed away during the flow along with debris,
- on the downstream of large floating debris,
- among closely spaced buildings.

2.7   Debris Modelling

An exact modelling of the tsunami-generated debris is difficult since there are different types of debris generated during a typical tsunami event. However, most of the researchers model the random shape of the debris as cylindrical or cuboid shape. Researchers consider a particular kind of prototype debris like containers or wooden logs and model it using Froude's law. During early stages of research, a prototypical wooden log was assumed and modelled as cylindrical or cuboid-shaped debris (Haehnel and Daly, 2004; Nouri *et al.*, 2009). It was observed during the Indian Ocean tsunami and the Tohoku tsunami 2011 that many shipping containers, boats, ships and barges were brought to the shore by the huge tsunami waves. Hence, a few researchers modelled containers as debris and carried out impact studies on the structures, bridges and columns of buildings (Stolle *et al.*, 2015, 2017; Ko *et al.*, 2015; Madurapperuma and Wijeyewickrema, 2012).

The model was a simple box modelled for size and weight. The mass of debris is varied by adding additional weights inside the debris so as to model the shipping containers in empty and loaded conditions. The debris can be modelled with acrylic sheets, aluminium or FRP. It is modelled such a way that it is rigid and stiff so there will not be much error in the measured impact force.

2.8 Physical Model Tests

The rapidly advancing tsunami bore closely resembles the hydraulic bore generated during a dam-break event (Chanson, 2005, 2006). Therefore, most of the previous studies used the dam-break bores setup to study the forces on the structures (Al-Faesly *et al.*, 2011; Shafiei *et al.*, 2016). However, the incoming tsunami bore can be non-breaking, breaking or an undular bore which depends on the intensity of the tsunami wave, origin point of tsunami and bathymetry of the shore. Bryant (2014) and Harish *et al.* (2018) carried out a series of experiments on the impact of debris on structures using undular bore generated in the laboratory. The undular bore moving towards the shore can be categorized as elongated single pulse, symmetrical N waves or unsymmetrical N waves. These different types of waves were generated in the laboratory using in-house WaveGen software, which was previously adopted for the large-scale experiments in the Large Wave Flume, Hannover (Schimmels *et al.*, 2016; Sriram *et al.*, 2016).

In case of hydraulic bore generated using dam break, the debris is picked up by the hydraulic bore and carried away along with the surging bore (Nouri *et al.*, 2009; Shafiei *et al.*, 2016). However, Harish *et al.* (2018) reported that the same phenomenon is not observed in the case of undular bore. The motion of the debris in undular bore is quite different. The motion of the debris in undular bore purely depends on the characteristics of the undular bore. It was reported that in elongated single pulse wave (Fig. 2.6(a)), the debris heaved predominantly with small surge motion. During a long period, undular waves (symmetrical or asymmetri-

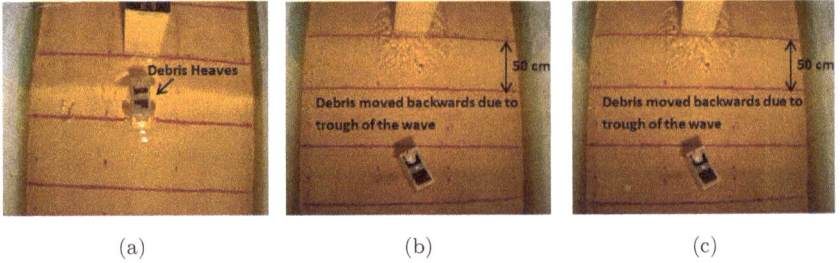

(a)   (b)   (c)

Fig. 2.6   Motion of debris in different wave forms.

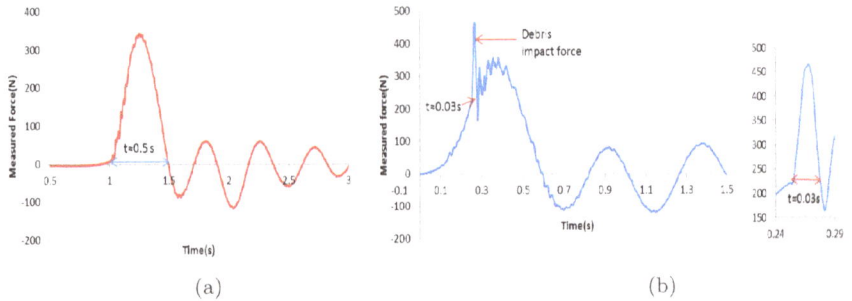

(a)   (b)

Fig. 2.7   Typical time history of the wave force (a) and debris impact force (b) acting on the structure for an unsymmetrical wave.

cal N waves) (Fig. 2.6(b)), along with the crest of the wave (surge motion) are observed. However, in undular bore of comparatively small time period (symmetrical or asymmetrical N waves) (Fig. 2.6(c)), the debris is carried away along with the surging bore (surge motion). The force acting on the structure primarily depends on the characteristics of the wave. The debris carried by the undular bore in a short period imparted large force on the structure, since the debris moved along with the moving wave as can be seen in Fig. 2.6.

2.8.1   *Necessity of considering debris impact*

In order to study the effect of debris impact on structure, Harish *et al.* (2018) conducted an experiment with and without debris conditions to understand the debris impact loading. From Fig. 2.7, it is observed that the debris imparts a short duration impulse on the structure. The debris impact force is quite large. The contact duration of the debris on structure is very less compared to the wave impact on structure alone. ASCE/SEI 7-05 suggests the contact duration of 0.03 s between the structure and the debris. Thus, it is necessary to include the debris impact force while designing category III and category IV structures.

2.9 Summary

It is well understood that the debris is the main source of large impact load on the structures during the ingress of a tsunami. Although it is not possible to design all the structures for debris impact load, at least category III and category IV structures located close to the coast need to be designed for the loading due to debris impact. This is one of the important lessons learnt from the huge mortality and monetary loss suffered. As an exact prediction of debris impact loading on structures during real tsunami is impossible, ASCE provides rational formulas that can be used as design debris load.

References

Al-Faesly, T., Nistor, I., Palermo, D., and Cornett, A. (2012). "Simulated tsunami bore impact on an onshore structure." Proceedings of the 20th Canadian Hydrotechnical conference, HY-025-01-HY-025-010.

ASCE/SEI 7-05 (2005). "Minimum design loads for buildings and other structures." ASCE Standard, 376.

ASCE/SEI 7-16 (2016). "Minimum design loads for buildings and other structures." ASCE Standard.

ASTM D2487-11 (2011). "Standard practice for classification of soils for engineering purposes (unified soil classification system)." ASTM International, West Conshohocken, PA.

Blair, T. C., and McPherson, J. G. (1999). "Grain-size and textural classification of coarse sedimentary particles." Journal of Sedimentary Research, 69(1), 6–19.

Bryant, E. (2014). Tsunami: The Underrated Hazard, Springer, Berlin, p. 330.

Chanson, H. (2005). "Analytical solution of dam-break wave with flow resistance: Application to tsunami surges." Proceedings of the 31st Biennial IAHR Congress, Seoul, Korea, 137, 3341–3353.

Chanson, H. (2006). "Tsunami surges on dry coastal plains: Application of dam break wave equations." Coastal Engineering Journal, 48(4), 355–370.

FEMA: Coastal Construction Manual (2011). "Principles and practices of planning, siting, designing, constructing, and maintaining residential buildings in coastal area (4th edition)." FEMA P-55, Federal Emergency Management Agency.

FEMA (2012). "Guidelines for design for structures for vertical evacuation from tsunamis." FEMA P646, Federal Emergency Management Agency, Washington, USA.

Goseberg, N., Nistor, I., and Stolle, J. (2017). "Tracking of 'Smart' debris location based on RFID technique." Proceedings of Coastal Structures & Solutions to Coastal Disasters Joint Conference.

Haehnel, R. B., and Daly, S. F. (2004). "Maximum impact force of woody debris on floodplain structures." *Journal of Hydraulic Engineering*, 130(2), 112–120.

Haehnel, R. B., and Daly, S. F. (2002). "Maximum impact force of woody debris on floodplain structures." Technical Report ERDC/CRREL TR-02-2, U.S. Army Corps of Engineers. Springfield, Virginia.

Harish, S., Sriram, V., Sundar, V., Sannasiraj, S. A., and Didenkulova, I. (2018). "Experimental investigation of floating debris impact loading on structures during extreme waves like tsunami." In the 28th International Ocean and Polar Engineering Conference, International Society of Offshore and Polar Engineers.

Ko, H. S., Cox, D. T., Riggs, H. R., and Naito, C. J. (2015). "Hydraulic experiments on impact forces from tsunami-driven debris." *Journal of Waterway, Port, Coastal, and Ocean Engineering*, 141(3), 04014043.

Madurapperuma, M. A. K. M., and Wijeyewickrema, A. C. (2012). "Inelastic dynamic analysis of an RC building impacted by a tsunami water-borne shipping container." *Journal of Earthquake and Tsunami*, 6(01), 1250001.

Matsutomi, H. (2009). "Method for estimating collision force of driftwood accompanying tsunami inundation flow." *Journal of Disaster Research*, 4(6), 435–440.

Matsutomi, H. (1999). "A practical formula for estimating impulsive force due to driftwoods and variation features of the impulsive force." *Journal of Hydro, Coastal and Environmental Engineering, JSCE, Japan*, 621(II-47), 111–127 (in Japanese).

Naito, C., Cercone, C., Riggs, H. R., and Cox, D. (2013). "Procedure for site assessment of the potential for tsunami debris impact." *Journal of Waterway, Port, Coastal, and Ocean Engineering*, 140(2), 223–232.

Nistor, I., Nouri, Y., Palermo, D., and Cornett, A. (2009). "Experimental investigation of the impact of a tsunami-induced bore on structures." *Coastal Engineering*, 2008(5), 3324–3336.

Nouri, Y., Nistor, I., Palermo, D., and Cornett, A. (2010). "Experimental investigation of tsunami impact on free standing structures." *Coastal Engineering Journal*, 52(01), 43–70.

Paris, R., Naylor, L. A., and Stephenson, W. J. (2011). "Boulders as a signature of storms on rock coasts." *Marine Geology*, 283, 1–11.

Peterson, K., and Naito, C. (2012). "Simplified ABAQUS model of ISO shipping container." Research Report, Lehigh University.

Schimmels, S., Sriram, V., and Didenkulova, I. (2016). "Tsunami generation in a large scale experimental facility." *Coastal Engineering*, 110, 32–41.

Shafiei, S., Melville. B. W., Shamseldin, A. Y., Adams. K. N., and Beskhyroun. S. (2016). "Experimental investigation of tsunami-borne debris impact on structures: Factors affecting impulse-momentum formula." *Ocean Engineering*, 127, 158–169.

Sheth, A., Sanyal, S., Jaiswal, A., and Gandhi, P. (2006). "Effects of the December 2004 Indian Ocean tsunami on the Indian mainland." *Earthquake Spectra*, 22, 435–473.

Sriram, V., Didenkulova, I., Sergeeva, A., and Schimmels, S. (2016). "Tsunami evolution and run-up in a large scale experimental facility." *Coastal Engineering*, 111, 1–12.

Stolle, J., Goseberg, N., Nistor, I., and Petriu, E. (2017). "Probabilistic investigation and risk assessment of debris transport in extreme hydrodynamic conditions." *Journal of Waterway, Port, Coastal, and Ocean Engineering*, 144(1), 04017039.

Stolle, J., Nistor, I., Goseberg, N., Matsuba, S., Nakamura, R., Mikami, T., and Shibayama, T. (2015). "Flood-induced debris dynamics over a horizontal surface." Proceedings of Coastal Structures and Solutions to Coastal Disasters.

Yeh, H., Barbosa, A. R., Ko, H., and Cawley, J. G. (2014). "Tsunami loadings on structures: Review and analysis." *Proceedings of Coastal Engineering*, 1(34), 4.

Chapter 3

Tsunami Hazards and Aspects on Design Loads

3.1 Introduction

A tsunami is generally characterized as shallow water waves in a period of a few hours and of speed greater than 600 kmph near the generation point. As it reaches shallow water depths, its speed and length reduce. Furthermore, as the period of the wave remains same, more water is forced between the wave crests causing the height of the wave to increase. Because of this *shoaling* effect, a tsunami that was unnoticeable in deep water might grow to wave heights of several metres. As stated in the earlier chapter, if the trough of the tsunami wave reaches the coast first, a phenomenon called *drawdown* occurs or in other words sea water recedes towards the ocean; however, very soon, its crest hits the coast.

The knowledge on characteristics of the tsunami, its generation, propagation and how it differs from the waves observed on the beach is of great importance in prediction and early warning. This is an important aspect to minimize the damage in future. Estimating the flooding area of the coastal zone caused by the tsunami waves is essential for tsunami hazard mitigation.

In order to determine run-up of long waves, different theoretical and experimental studies have been performed. The Federal Emergency Management Agency's (FEMA) Coastal Construction Manual (2005) and the City and County of Honolulu Building Code (CCH) (2000) provide design guidelines for buildings exposed to tsunamis. Okada *et al.* (2004a and 2004b) provided design guidelines for tsunami refuge buildings based on the estimation of tsunami loads from the laboratory study (Asakura *et al.*, 2000).

3.2 Characteristics of Tsunami

3.2.1 *General*

In this chapter, let us consider the perspectives of tsunami only in terms of its impact on the coasts. The tsunami wave propagates from its source of generation in all possible directions. The major wave propagation direction is dictated by the false plane (in the case of earthquake), i.e., the generated wave crest is parallel to the false plane. For instance, in the 2004 Indian Ocean tsunami, the fault plane was about 1,000 km in the north–south direction and about 200 km in the east–west direction. Hence, a large wave propagates along the east–west direction.

Similar to the definition of wind waves, the basic tsunami wave characteristics are wave length and wave period. However, these two characteristics are only related to water depth in the case of tsunami as discussed earlier, i.e., it simply follows a shallow water wave propagation case due to the relatively larger wave length compared to water depth. In other words, the speed of tsunami and the dispersive nature of the tsunami are mainly constrained by the water depth. On the landfall point, the two main characteristics related to tsunami coastal impact are run-up and inundation.

3.2.2   *Tsunami landfall characteristics*

Three basic landfall characteristics of tsunami are run-up, inundation and time of landfall. The last one becomes important only when multiple waves sweep the coast. In other cases, it is of importance only for hydro-dynamists to understand the different wave propagation patterns.

3.2.2.1   *Run-up*

The vertical height of maximum water surface elevation at the tsunami landfall point is called a run-up at that particular location of the coast. The datum in general is mean sea level or chart datum. It varies over a wider range even within say a stretch of about 10 km. The sea level rises, which is the inundation height or run-up, R, and the distance up to which the tsunami runs into the land is termed as the inundation distance, L, as shown in Fig. 3.1. The run-up due to the tsunami depends on near-shore shallow-water bottom topography, the tsunami approach angle and any obstruction such as an island in front of the coast under consideration. In most of the calculations, it is simple to assume that a tsunami is a shallow-water wave. Analytical solution of the shallow-water equations for the prediction of long wave run-up over smooth plane beaches can be derived.

The run-up of non-breaking long waves was derived theoretically and obtained the following simple power law for the prediction of solitary wave

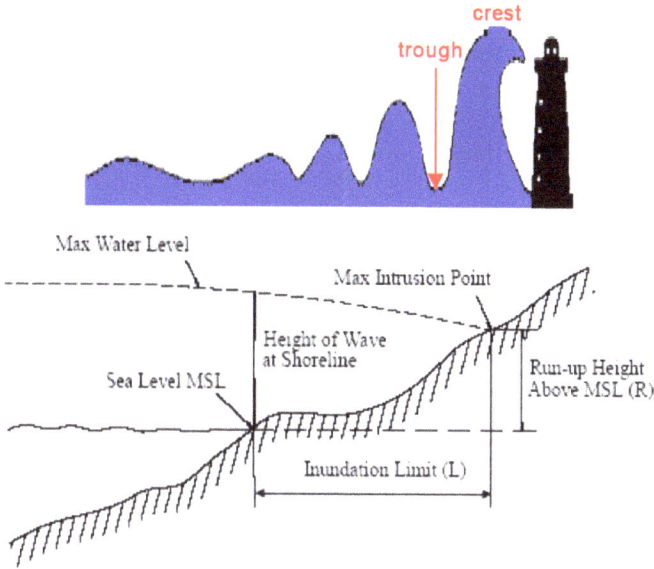

Fig. 3.1   Definition inundation height and distance.

run-up on a smooth plane beach:

$$\frac{R}{d} = 2.831 \sqrt{\cot \beta} \left(\frac{H}{d}\right)^{5/4}, \tag{3.1}$$

where R is the run-up, H is the wave height, d is the water depth and β is the inclination angle of the plane beach. The run-up is defined at the landfall point. However, the water depth in the calculation of run-up needs a definite value. A water depth of 10 m is generally adopted to estimate the run-up. After the tsunami makes a landfall, the run-up is estimated from the local maxima along the horizontal inundation stretch.

3.2.2.2   *Inundation*

The other important terminology in tsunami impact on coastal zone is inundation. It is the horizontal distance over which the sea spreads its tongue during the tsunami landfall. It is the function of run-up and the onshore land topography. The coastal morphological features of the particular shoreline play a major role in dictating the inundation levels. However, if there is a creek, river mouth or backwaters at the coast, the impact due to inundation will be more severe due to rush up of water through the weak openings at the coast. It causes inundation from both sides of land sandwiched between ocean and backwaters.

3.3   Tsunami Hazards

3.3.1   *General*

The tsunami imposes so-called hazards only on the coastal zone and near-shore waters. Its influence is generally unnoticed on a ship in deep waters. The major water circulation in deep water might affect the atmospheric processes and then have an impact in the climatology such as change in monsoon pattern and cyclonic systems. There are three major effects of a tsunami: inundation, wave impact on structures and scour erosion. The tsunami-induced hazards can be generally categorized into impacts on coastal zone, influence on the near-shore sailing vessels and impact inside the ports.

3.3.1.1   *Impacts on coastal zone*

The effect of tsunami on the coasts can be broadly classified under four major categories, namely flotation, impact, receding and submergence. During the initial tsunami run-up towards the coast, the movable and loose objects such as cars, boats and people, can be dragged by the water mass by the way of flotation. Fishing boats and other small vessels are usually found to be dragged a few kilometres upstream (along backwaters/river course) and many of them will be left on bridges, jetties and elevated bunds. Considerable damage will also be caused by the floating debris, including boats and cars, that becomes dangerous projectiles after it crashes into buildings and breaks power lines. The flotation of navigational vessels also damages the berthing structures, particularly the top coping beam and pile heads are dented.

Second, the mass movement of water towards the coast results in an impact force on the shore-based structures. Strong, tsunami-induced currents lead to the erosion of foundations and the collapse of bridges and structures, in particular, seawalls. Even heavy objects such as overhead cranes on a berth inside a harbour or a vertical wall either a masonry wall as a component of residential building or a shore-protection structure are subjected to heavy damage. The impact force also destabilizes the floating objects resulting in mass destruction along with the effect of flotation. Flotation and drag forces move houses and overturn railroad cars.

Considerable damage is caused by the resultant floating debris, including boats and cars that become dangerous projectiles that may crash into buildings, break power lines and may start fires. Of increasing concern is the potential effect of tsunami drawdown, when receding waters uncover cooling water intakes of nuclear power plants. The acceleration of the moving water mass into the land decelerated quickly due to the elevation of the land. The quick deceleration of the moving water induced the movement of the water in the opposite direction towards the sea. The reversal in the flow of water towards the sea has also been accelerated by the natural slope of the land. Hence, the wave had the tremendous power to drag away any loose material on its way towards the sea, which forms the third category of the devastation. One of the severe impacts during the receding of water is the scour around foundations and trees. Strong, tsunami-induced currents lead to the erosion of foundations leading to the collapse of bridges and seawalls. The houses are mostly damaged by foundation failures as a result of their complete exposure.

Finally, the fourth category is the damage aggravated due to the inundated water which could recede to the sea immediately. The inundated seawater in the lake and ponds would affect the ground water, and most of the cultivable land near the coast could be polluted due to seawater intrusion. The long duration of the inundation causes most deaths due to the spread of diseases, power outages and machinery malfunction.

3.3.1.2 *Impact on ports*

Once a long wave such as a tsunami enters into a partially closed basin such as harbours with a narrow entrance, the water surface persistently oscillates for many hours/days without losing its energy. Fire from damaged ships in ports or from ruptured coastal oil storage tanks and refinery facilities can cause damage greater than that inflicted directly by the tsunami. Boats that were neatly lined up in their moorings along three piers of Chennai port as seen in the satellite image of 14 August are completely disturbed and some of them were seen on top of the piers soon after tsunami through the image of 29 December 2004 (Fig. 3.2). The impact on the tsunami on the ports severely affect its functionality and efficiency apart from the damage to the cargo as well as to the port facilities.

In addition, if the wave length of one such long wave has been of the harmonic order of basin's characteristic length, then so-called resonance would occur, known as harbour resonance. The chance of occurrence of resonance is high since once trapped, the long wave forms into many wave components

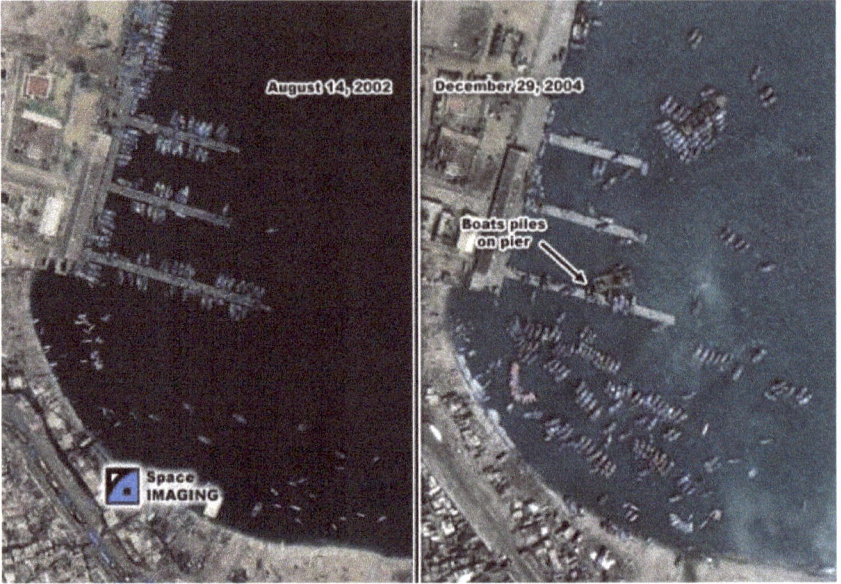

Fig. 3.2   A view on the effect of tsunami inside a harbour.

in the range of 10 min to an hour. The corresponding wave length is of the
order of few kilometres to several tens of kilometres depending on the water
depth in the basin. Hence, the wave length might form in terms of multiples of basin length/width. Once the resonance is set, the wave amplitude
increases multifold unless there is a strong dampening mechanism. Most
of the artificial harbours protected with rubble mound breakwaters have
natural dampening system in terms of absorbing face of the breakwaters.

The harbour oscillation inside the basin thus depends on the following:

(a) the tsunami wave period as it enters into the basin;
(b) the characteristic dimensions of the harbour basin;
(c) absorption and reflection characteristics of basin boundaries dictated
    by the type of breakwater boundaries and berth face.

The resonant period of oscillation (T) inside a harbour basin with characteristic length (L) and breadth (B) is given as follows:

$$T = \frac{2}{\sqrt{gd}} \left[\left(\frac{n}{L}\right)^2 + \left(\frac{m}{B}\right)^2 \right]^{\left(-\frac{1}{2}\right)}$$

$$n = 0, 1, 2, 3 \ldots \text{and} \quad m = 0, 1, 2, 3.$$

(3.2)

Here, n and m are integer numbers representing each mode of oscillation following the definition of Raichlen (1966). For the characteristic dimension of the Chennai harbour basin with $d = 15$ m and $L = 1000$ m, a resonant tsunami period for $n = 1$ and $m = 0$ is of the order of 3 min. The first mode oscillation is severer than higher modes ($n > 1$). The above equation is valid in a closed basin. Depending on the width opening at the harbour entrance, the aforementioned equation should be modified.

3.4 Tsunami Forces on Coastal Structures

3.4.1 *General*

Large bodies such as vertical-faced walls need to be designed considering hydrostatic force, hydrodynamic force and impact force due to waves and floating objects. The buoyant force and static pressure head on the sea and lee sides of the structure form under hydrostatic pressure head. For small water piercing structures such as pile supported jetty, wharves, etc., only the hydrodynamic force and impact force are to be considered.

3.4.2 *Possible solutions to consider tsunami effects on structures*

A list of tsunami effects on the structures and the possible solutions as suggested by NOAA (2001) is presented in Table. 8.1 which serves as a ready reference.

3.4.3 *Hydrostatic force*

The hydrostatic force can be estimated as follows:

$$\left. \begin{array}{l} F_h = \dfrac{1}{2}\rho g h\,(2d + 6\eta - h) + \dfrac{u_t^2 h_1}{2g} \quad \text{for } h_1 = h \\[2mm] F_h = \dfrac{1}{2}\rho g h_1^2 + \dfrac{u_t^2 h_1}{2g} \quad \text{for } h_1 < h \\[2mm] h_1 = \min(h,\, d + 3\eta) \end{array} \right\} \tag{3.3}$$

where ρ is the seawater density; g is the gravitational acceleration; h is the height of the structure; d is the water depth at the toe of the structure; η is the tsunami wave height at 10 m water depth offshore; u_t is the water particle velocity which can be assumed to be equal to the maximum tsunami flow velocity which is equal to $2\sqrt{g(d + \eta)}$. The tsunami run-up at the structure is adopted as three times the tsunami wave height at 10 m

Table 3.1.   Tsunami effects and design solutions (NOAA, 2001).

Phenomenon	Effect	Design solution

Inundation	Flooded basements
	Flooding of lower floors
	Fouling of mechanical,
	  electrical, and communication
	  systems and equipment
	Damage to building materials,
	  furnishings and contents
	  (supplies, inventories,
	  personal property)
	Contamination of affected area
	  with Water-borne pollutants
		Choose sites at higher elevations
		Raise the building above the
		  flood elevation
		Do not store or install vital
		  material and equipment on
		  floors or basements lying
		  below tsunami inundation
		  level
		Protect hazardous material
		  storage facilities that must
		  remain in tsunami hazard
		  areas
		Locate mechanical systems and
		  equipment at higher locations
		  in the building
		Use concrete and steel for
		  portions of the building
		  subject to inundation
		Evaluate bearing capacity of
		  soil in a saturated condition

	Hydrostatic forces (pressure on
	  walls caused by variations in
	  water depth on opposite
	  sides)
		Elevate buildings above flood
		  level
		Anchor buildings to foundations
		Provide adequate openings to
		  allow water to reach equal
		  heights inside and outside of
		  the buildings
		Design for static water pressure
		  on walls

	Buoyancy (flotation or uplift
	  forces caused by buoyancy)
	Saturation of soil causing slope
	  instability and/or loss of
	  bearing capacity
		Elevate buildings
		Anchor buildings to foundations
		Evaluate bearing capacity and
		  shear strength of soils that
		  support building foundations
		  and embankment slopes
		  under conditions of saturation
		Avoid slopes or provide setback
		  from slopes that may be
		  destabilized when inundated

Currents	Hydrodynamic forces (pushing
	  forces caused by the leading
	  edge of the wave on the
	  building and the drag caused
	  by flow around the building
	  and overturning forces that
	  result)
		Elevate buildings
		Design for dynamic water forces
		  on walls and building
		  elements
		Anchor building to foundations

Table 3.1.   (*Continued*)

Phenomenon	Effect	Design solution
	Debris impact	Elevate buildings
		Design for impact loads
	Scour	Use deep piles or piers
		Protect against scour around foundations
Wave break and bore	Hydrodynamic forces	Design for breaking wave forces
	Debris impact	Elevate buildings
		Design for impact loads
	Scour	Design for scour and erosion of the soil around foundations and piles
Drawdown	Embankment instability	Design water front walls and bulkheads to resist saturated soils without water in front
		Provide adequate drainage
	Scour	Design for scour and erosion of the soil around foundations and piles
Fire	Water-borne flammable materials and ignition sources in buildings	Use fire-resistant materials Locate flammable material storage outside of high hazard areas

water depth offshore as shown in Fig. 3.3. The static pressure force due to flow velocity head is adopted to be uniform throughout the water depth. However, for transparent structures such as open-pile type structures, the hydrostatic force can be neglected.

Nakano and Paku (2005) have conducted extensive surveys in the areas of Sri Lanka and Thailand that were damaged by the tsunami generated by the Sumatra earthquake. Data were collected on damage found on tens of coastal structures that had little influence on the presence of obstructions, including buildings, RC pillars, brick walls and elevated water reservoirs. Using these data, the validity of the design equation for tsunami load was examined. The value of α, that is the dividing line between damage and no damage, was calculated separately for wall members and column members.

$$p_x = \rho g(\alpha \eta_{\max} - z), \tag{3.4}$$

where p_x is the maximum tsunami wave pressure ($0 \leq z/\eta_{\max} \leq 3$). z is the height of the relevant portion from ground level. $\eta_{\max}$ is the maximum inundation depth.

Calculating the value of α in tens of examples of damage revealed that wall members were undamaged roughly when $\alpha > 2.5$; therefore, the value

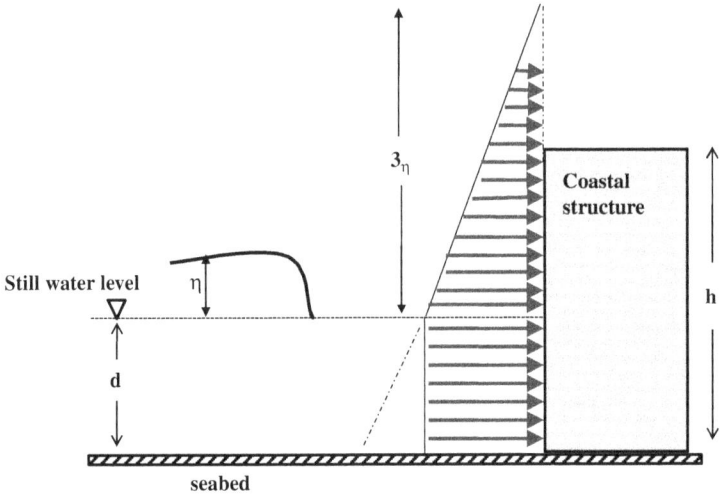

Fig. 3.3   Tsunami incidence on the coastal structure.

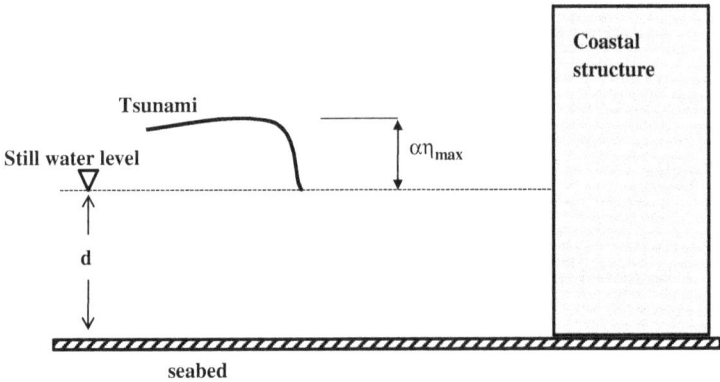

Fig. 3.4   Experiment setup.

$\alpha = 3.0$ for the coefficient in the tsunami wave pressure equation is considered appropriate. $\alpha = 2$ is found to be the dividing line between the damaged and undamaged conditions of the column members.

Further, the tsunami induced wave pressure is formulated based on the experiment as shown in Fig. 3.4 by Ikeno *et al.* (2001). The water pressure is measured by bore wave level of the flush tank divided at the gate.

$$p_m(z) = 2.2\alpha\,\rho g\,(\eta_{\max} - z/3), \tag{3.5}$$

where p_m is the maximum tsunami wave pressure ($0 \leq z/\eta_{max} \leq 3$: above the surface of still water), η_{max}, tsunami wave amplitude and α, the Impact Coefficient (=1.36).

3.4.3.1 Leeside water pressure while overtopping

During the tsunami inundation on the lee side of seawalls, the pore water pressure increases due to the saturation of the soil. This pore water pressure might exist for a longer period which would lead to destabilization of the structure. Hence, provision of pore pressure release system in wall type structures and drains in seawalls is recommended.

3.4.3.2 Buoyant force

The submergence of structure due to the rise in water level reduces the effective weight of the structural elements which in other words, reduces the stiffness of the structure. This can be adopted in terms of upward force, i.e., buoyant force, F_B.

$$F_B = \rho g V, \tag{3.6}$$

where V is the submerged volume of structural elements.

3.4.3.3 Hydrodynamic wave force

Due to the nature of the long tsunami wave, drag force is induced on relatively small characteristic dimension of the structural elements. The hydrodynamic drag force, F_d can be estimated from the following:

$$F_d = \frac{1}{2}\rho C_d A u_t^2, \tag{3.7}$$

where C_d is the drag coefficient = 1.2 for circular structures; 2.0 for square buildings; 1.5 for wall sections where the length of the wall is less than wave length. A is the projected area of the elements on the plane normal to the flow direction. Equation (3.7) is valid if the structure dimension (B) is small relative to tsunami wave length (L), i.e., $B \ll L$.

On the other hand, the tsunami wave force (F_H) is represented as the combination of drag, inertia, impulse and hydraulic gradient forces by Omori et $al.$ (2000).

$$F_H = \frac{1}{2}\rho C_D u|u|B\eta + \rho C_M \dot{u} BL\eta + \frac{1}{2}\rho C_S(\theta)u|u|B\eta + \rho g BL\eta\frac{d\eta}{dx}, \tag{3.8}$$

where B: width of the structure, L: length of the structure, C_D: drag coefficient ($= 2.05$), C_M: mass coefficient ($= 2.19$), $C_S(\theta)$: impact coefficient ($= 3.6 \tan\theta$), Θ: angle of wave, u: velocity of tsunami wave in the direction of advance, $\dot{u}$: acceleration of tsunami wave in the direction of advance and η: inundation depth.

3.4.4   Tsunami force on inland structure

Iilzuka and Matsutomi (2000) proposed the equation to estimate the hydrodynamic tsunami force (F_{HD}) based on the damage of houses while tsunami hit in Japan. The equation is formulated in order to make the relationship between drag forces and damage degrees of houses.

$$F_{HD} = \frac{1}{2}\rho C_D u^2 h_f B_h \tag{3.9}$$

where C_D: drag coefficient ($= 1.1 \approx 2.0$), u: velocity in inland, h_f: inundation depth in front of the structure and B_h: width of structure.

3.4.5   Impact force

The impact force needs to be considered for any structure elements facing the water front. The impact is caused not only due to impulsive wave but also due to the floating objects. Tsunamis often approach the shore in the formation of bores depending on the bottom bathymetry. The impulsive force can be reduced if the structure is elevated. The pressure impulse is an integrated quantity as a function of maximum wave impact and the duration of its attack. The details of the impact force over elevated structures are discussed in Section 12.3.4 of this book. The impact force due to a floating object is equal to the mass of the probable floating objects times the acceleration induced into the floating objects by the tsunami flow velocity. In a harbour, the severe impact may be due to the floatation of boats and ships.

The impact force, F_I can be assumed to act at still water level, i.e., at $(h + 3\eta)$ above the seabed level where tsunami landfall occurs.

$$F_I = m\frac{du_b}{dt} = m\frac{u_t}{\Delta t}, \tag{3.10}$$

where u_b is the velocity of probably floating object and during the tsunami wave propagation, the floating body is assumed to activate from its rest to the flow velocity, u_t ($du_b = u_t$). Δt is the impact duration which is the time between initial contact of the floating body with the building and

Table 3.2.   Recommended scour depth in front of the obstruction.

Soil type	Distance from shoreline as function depth, d	
	100 m	>100 m
Loose sand	0.8 d	0.6 d
Dense sand	0.5 d	0.35 d
Soft silt	0.5 d	0.25 d
Stiff silt	0.25 d	0.15 d
Soft clay	0.25 d	0.15 d
Stiff clay	0.1 d	0.05 d

the maximum impact force. This impact duration depends on the natural frequency of the structure. This value can be 1.0 s for wooden structure, 0.5 s for steel structure and 0.1 s for reinforced concrete structure. The details of the impact force prediction on the structures due to the impulse-type loading are discussed in detail in Chapter 12.

3.4.6   *Scour depth*

One of the severest devastation effects of the tsunami wave is the scour under the foot of the structure while the tsunami wave receded with a speed proportional to the slope of the land topography and the tsunami height. The effect of scour exposes the free length of the coastal structures which makes say the embedment length of the pile become less than the design requirement. The caisson-type block structural elements can be destabilized with the scour on the seaside and the excess water pressure on the land side of the block. The scour depth for coast-based structures such as berthing structures should be adopted to fix the founding depth of the structure. Table 3.2 presents an estimate of maximum scour as a function of water depth level in front of the pile/structure, d, for different soil conditions. The scour depth may be suitably reduced if there is a first level defence in front of the structure such as vegetation, sand dune, breakwater, etc. It should be noted that the Chennai harbour basin got deepened on an average by 2 m during the 2004 tsunami. Such measurement for the loose silty sand is less than given in Table 3.2 due to the presence of breakwaters.

3.5   Summary

The tsunami impact on coastal zone has been analysed based on two major concerns: (i) Saving lives/reducing human suffering and (ii) limiting economic losses/promoting sustainable development. It is well understood that the tsunami disaster cannot be prevented and needs preparedness in advance as the warning provided will be too short a time to respond. It is our task to reduce its effect. An integrated multi-hazard approach with an emphasis on natural coastal hazards like storm surge, cyclone and tsunami is an essential task before us. The infrequent low probable tsunami hazard is coupled with frequent cyclonic hazard to justify the investments. The specific engineering measures ranges from the construction of elevated shelters for the safety of people to the plantation of mangroves/coastal forests for reducing the impact on the coastal zone. Wherever possible/required, the hard measures of coastal protection such as seawalls or coral reefs and harbour-like protection using breakwaters can be planned. A detailed environmental impact study is an essential task before implementing any hard measures. As the occurrence of a tsunami is a rare event, the day-to-day operational activities such as fishing, tourism should not be affected. The bio-shields or plantation can be considered with due respect on coastal morphology as a long-term solution. A revised planning strategy may be adopted for identified vulnerable structures. This is an essential task for critical infrastructures such as power stations, warehouses, oil and other storages tanks, etc. Wherever possible, retrofitting of such structures can be implemented.

References

Asakura, R., Iwase, K., Ikeya, T., Takao, M., Fujii, N., and Omori, M. (2000). "An experimental study on wave force acting on on-shore structures due to overflowing tsunamis." *Proceedings of Coastal Engineering, Japan Society of Civil Engineering*, 47, 911–915.

Bryant, E. (2001). *Tsunami: The Underrated Hazard.* Cambridge University Press, Cambridge.

City and County of Honolulu Building Code (CCH) (2000). "Department of planning and permitting of Honolulu Hawaii." Chapter 16, Article 11, Honolulu, Hawaii.

FEMA Coastal Construction Manual (2005). *FEMA 55 Report*, Edition 3, Federal Emergency Management Agency, Washington D.C.

Iilzuka, H., and Matsutomi, H. (2000). "Damage prediction of a tsunami flood." *Proceedings of the Coastal Engineering of JSCE*, 47, 381–385.

Ikeno, M., Mori, N., and Tanaka, H. (2001). "Experimental research about wave force of a bore and breaking wave and the action and the impulsive force of drifting timber." *Proceedings of the Coastal Engineerings of JSCE*, 48, 846–850.

Kurian, N. P., Rajith, K., Muralikrishnan, B. T., Kalaiarasan, P., and Abhilash Pillai, P. (2005). "December 2004 tsunami: Runup level and impact along the Kerala coast" in S. M. Ramaswamy and C. J. Kumanan (eds.), *Tsunami: The India Context*. Allied Publishers Private Ltd., Chennai.

Nakano, Y., and Paku, C. (2005). "Studies on lateral resistance of structures and tsunami load caused by the 2004 Sumatra earthquake Part-1 Outline of Survey." Summaries of technical papers of Annual Meeting Architectural Institute of Japan (Kinki).

NOAA (2001). "Designing for tsunamis: Background papers." U.S. Department of Commerce and the National Oceanic and Atmospheric Administration (NOAA).

NOAA (2001). "Seven principles for planning and designing for tsunami hazards." National Oceanic and Atmospheric Administration (NOAA), March.

Okada, T., Sugano, T., Ishikawa, T., Ogi, T., Takai, S. and Hamabe, T. (2004a). "Structural Design Method of Building to Seismic Sea Wave, No. 1 Preparatory Examination." Building Letter, The Building Center of Japan (in Japanese).

Okada, T., Sugano, T., Ishikawa, T., Ogi, T., Takai, S. and Hamabe, T. (2004b). "Structural Design Method of Building to Seismic Sea Wave, No. 2 Design Method (a Draft)." Building Letter, The Building Center of Japan (in Japanese).

Omori, M., Fujii, N., and Kyotani, O. (2000). "The numerical computation of the water level, the flow velocity and the wave force of the tsunami which over flow the perpendicular revetments." *Proceedings of the Coastal Engineering of JSCE*, 47, 376–380.

Raichlen, F. (1966). "Harbor resonance." *Estuary and Coastline Hydrodynamics*, A. T. Ippen, ed., McGraw-Hill, New York, 281–340.

PART 2

Field Studies

Chapter 4

Behaviour of Shoreline between Groyne Field and Its Effect on the Tsunami Propagation

4.1 Background

The stretch of the coast north of the port of Chennai, situated along the southeast coast of India, which experienced the attack of the mighty tsunami in 2004, has been experiencing continuous erosion since the late 1960s. A number of temporary measures and a seawall as a permanent solution to counter this problem have failed. This is an area of active development. In order to combat further erosion, a detailed study of this area was taken to suggest a suitable permanent coastal protection measure. The proposed coastal protection scheme consists of shore-connected groynes. The chapter presents the salient results on the shoreline evolution due to the proposed construction of groynes. The construction of the groyne field started in May 2004. The behaviour of the shoreline before and after the tsunami has been monitored, the results of which are discussed herein.

The port of Chennai is located along the east coast of the Indian peninsula. The perennial problems along the east coast of India in general have been the closure of river mouths, siltation of approach channels of the major ports at Chennai, Visakhapatnam and Paradip and erosion of the significant portion of the land mass on the north of the above ports. Ever since the formation of the harbour in Chennai port with breakwaters, the coast on its north has been subjected to erosion at a rate of about 8 m per year due to the predominant northerly drift. A part of the existing National Highway and the residential area nearer to this coastline have already been sacrificed to the sea. In spite of the provision of a seawall, the erosion continued in few pockets along the coast and the severely affected zone is shown in Fig. 4.1. The locations of the study area as well as another port in Ennore, north of the Chennai port, commissioned a few years ago are also indicated in Fig. 4.1.

The measures adopted over the past four decades did not solve the problem of coastal erosion. A highway is not only the main link between

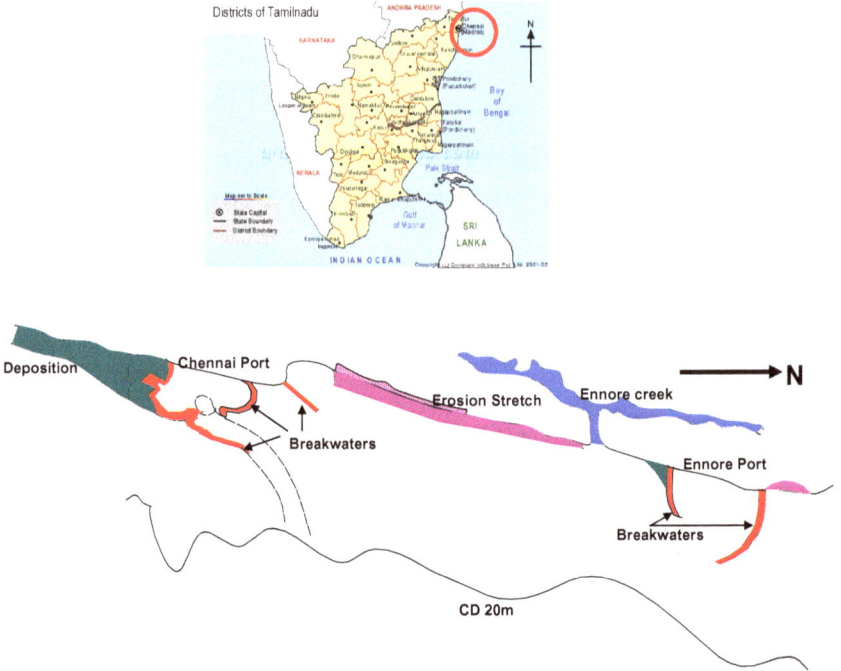

Fig. 4.1   Schematic view of the affected area in between Chennai and Ennore ports.

the ports of Chennai and Ennore but also serves as the only link to several industries along the coastal city that has been experiencing severe traffic congestion, many times to a standstill for hours together mainly due to the sacrifice of a part of the highway to the Bay of Bengal, which explains the importance of protecting this stretch of the coast. The solution of a groyne field to combat the erosion problem and the success of the project are highlighted herein.

4.2   Solution to Coastal Erosion

The solution for the coastal erosion problem was divided into two categories, a temporary strengthening of the existing seawall and a permanent remedial measure by providing suitable groynes. In the first phase, a detailed bathymetry survey for the measurement of an existing cross-section of seawall and its status in order to assess its adequacy for the design wave climate were carried out. The wave characteristics such as average wave height, wave

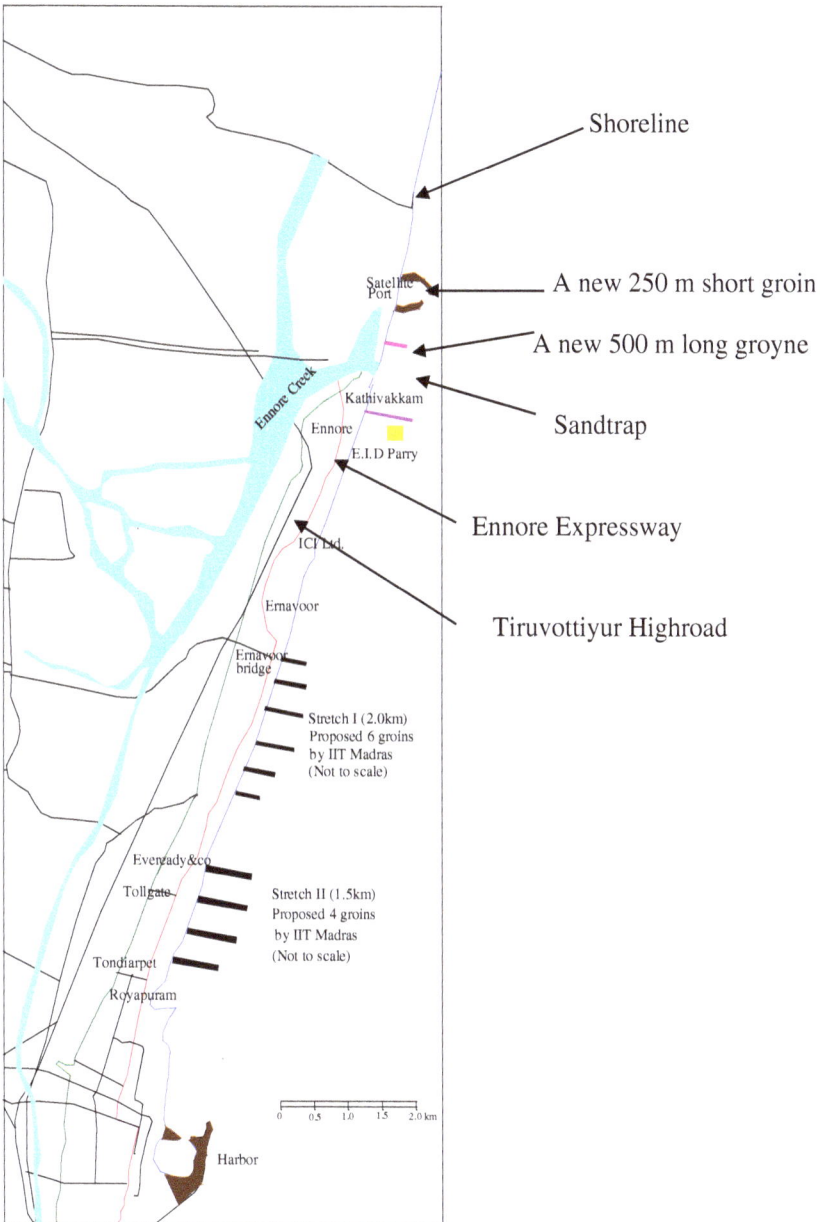

Fig. 4.2   Layout of the study area.

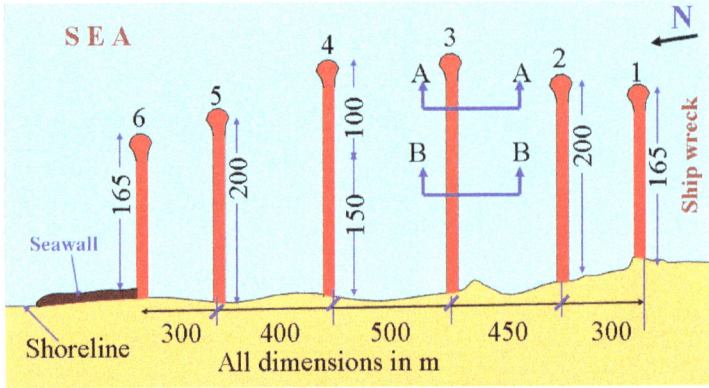

Fig. 4.3   Layout of the groyne field for stretch I.

period and wave direction, from which the average breaking wave characteristics were derived from the wave atlas, were prepared for the Indian
coast by National Institute of Oceanography (NIO). The monthly sediment transport has been estimated based on Energy Flux method (CERC,
1984), the method of Komar (1969) and by integrating the distribution
of sediments within the surf zone as suggested by Komar (1969). The net
sediment drift along the Chennai coast is observed to be about 0.8×10^6
m^3/year towards the North. In the second phase, as a permanent solution for the coastal erosion problem, 10 shore-connected straight rubble
mound groynes in the two severely affected stretches (stretch I and II)
shown in Fig. 4.2 were proposed. The length and the spacing between
groynes were designed based on the recommendations of Shore Protection
Manual (CERC, 1984), the details of which for stretch 1 alone is shown
in Fig. 4.3.

4.3   Response of Shoreline to Tsunami

In May 2004, the construction of the first phase of the groyne field commenced soon after which shoreline advancement on the south of the structure was witnessed. The groyne field for stretch 1 was completed prior
to the Indian Ocean tsunami in December 2004. In order to have a clear
picture on the effect of the tsunami on the shoreline variations, the measured beach widths formed only in between groynes 6 and 5 are shown in
Fig. 4.4. Similar results were seen in between the other groynes. The area of
the beach obtained through continuous monitoring for the different periods

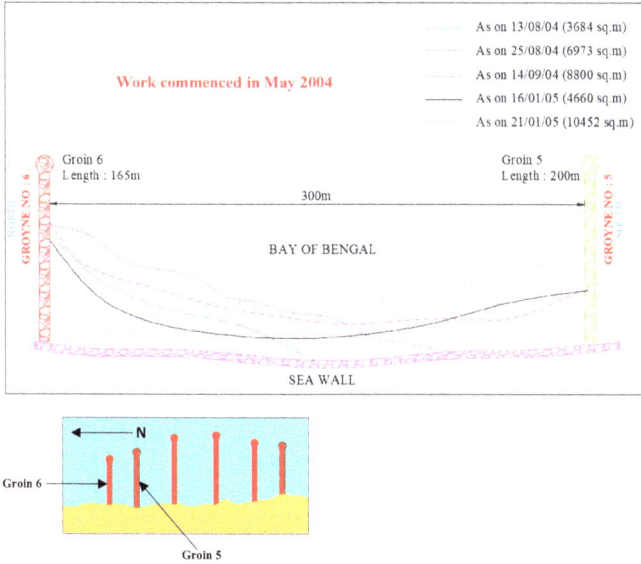

Fig. 4.4   Shoreline advance in between groynes 5 and 6 for different periods.

Table 4.1.   Area of beach in between groynes 5 and 6.

Date of measurement	Area in m^2
Work commenced in May 2004	
13 August 2004	3,700
25 August 2004	6,970
14 September 2004	8,800
Post-tsunami	
06 January 2005	4,660
21 January 2005	10,450

is shown in Table 4.1. The existing groyne field resisted the dynamics of the tsunami waves and also acted as a barrier to prevent excessive inundation and damage on the landward side of the coastline. The results clearly indicate the effect of the tsunami on the oscillation of the shoreline in between a pair of groynes. Soon after the groynes were constructed an advancement of the shoreline to an area of about 8,800 sqm was formed within a period of five months. The ingress of tsunami had taken away about 50% of the deposited sediments into the ocean which probably would have remained

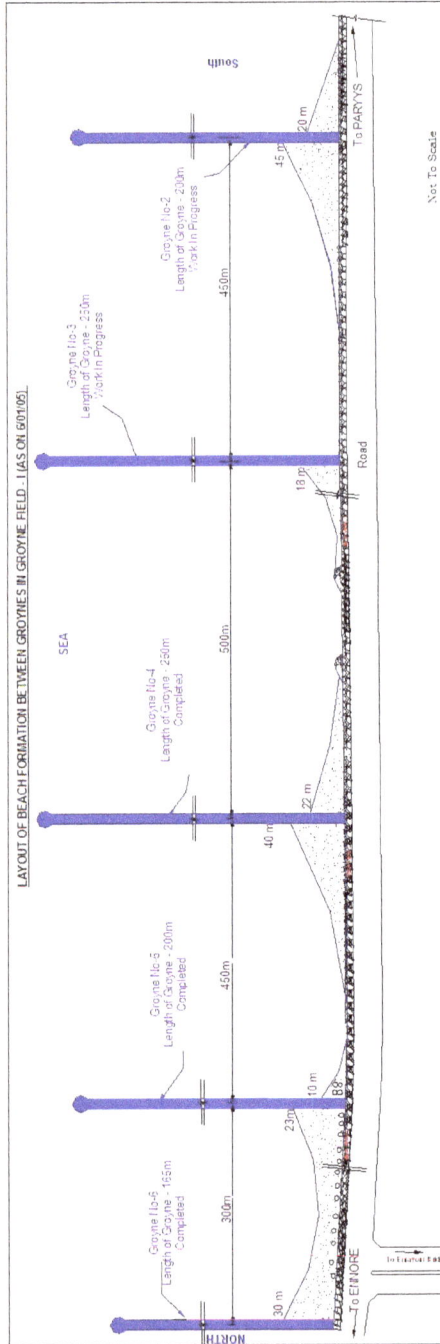

Fig. 4.5   The shoreline evolution due to the groyne field.

as an offshore bar, which was probably driven by the gravity waves and re-deposited, leading to an increase in the area of the beach by about 20%. Such a phenomenon has never been reported in literature. The evolution of the shoreline due to the entire groyne field for stretch I (post-tsunami) is projected in Fig. 4.5.

4.4   Summary

It is concluded that a perennial problem of coastal erosion swallowing a significant portion of the road connecting Chennai and Ennore ports, which also happens to be a national highway, has been solved. The solution was arrived at after careful and critical investigation that included planning of the protection measure, numerical modelling to assess its effectiveness, design of the groyne field, individual groynes, bathymetry survey, estimates and sequence of the construction. The sequence of construction of the groynes was carefully planned taking into account the seasonal variation in the direction of littoral drift at the Chennai coast. The tsunami run-up had initially taken away nearly 50% of the beach that was formed in between the groynes, and within a short duration of time slightly more than what was removed has been re-deposited.

References

CERC (1984). "Shore Protection Manual (SPM)." Vols. 1 and 2. U.S. Army Corps of Engineers, Vicksberg.

Komar, P. D. (1975). "Longshore currents and sand transport on beaches." *Proceedings of Civil Engineering in the Oceans/III*, 1, 333–354.

Chapter 5

Signature Studies (Tamil Nadu, Kerala, Andaman and Nicobar Islands)

5.1 General

Post-tsunami field surveys of the 2004 Indian Ocean tsunami have been conducted along the coast of the two maritime states of the mainland of the Indian peninsula, Tamil Nadu along the east coast and Kerala on the southwest. Surveys were also carried out along the coast of Andaman and Nicobar Islands, where the devastating effect of the tsunami was experienced. The study concentrated on the propagation of tsunami in terms of its time of arrival, run-up height and the inundation level, which depend on its geographic location, bathymetry and the tsunami characteristics.

5.2 Field Data Collection

5.2.1 *General*

Signature of the tsunami waves was captured at the earliest possible time after the occurrence of the event. In addition, secondary signatures, such as water-level marks on trees and buildings, usually difficult to identify, are the ones carrying information of the maximum height of the monstrous wave, were carried out. As this study had commenced soon after the tsunami, important data were successfully retrieved and recorded. The studies also collected information about changes in the ground level that had taken place around the affected areas.

The tsunami waves generated over the rupture zone as detailed earlier struck the Indian coast at different times with different intensities. The variation in the arrival time of the tsunami waves at different locations is due to the proximity of the coast from the source region, the variation in the sea bottom topography as well as the variation in the originated time of tsunami along the fault line that had occurred over the period of time. The post-tsunami survey along the entire Tamil Nadu coast from Kanyakumari district to north of Chennai was conducted from 10 to 22 January 2005. The instruments used were auto level, GPS and audio/video recorders for recording while interviewing the local people.

The signature studies for Andaman and Nicobar Islands were carried out starting from the Port Blair area on 21 January 2005 and continued for about six days. The team travelled to distant places, namely Chidiya Tapu, Wandoor, Havelock Island, Neil Island, Mayabunder and Diglipur. Estimation of run-up levels in these areas was carried out with the support from Andaman and Lakshadweep Harbor Works (ALHW) staff and Forest Department staff on duty. In the south Andaman Islands and Nicobar group of Islands, the run-up estimates were based on interviews with ALHW staff. The study areas of Tamil Nadu, Andaman and Nicobar Islands and Kerala are projected in Figs. 5.1(a), 5.1(b) and 5.1(c), respectively.

5.2.2   *Andaman and Nicobar Islands*

Andaman and Nicobar Islands were the most affected territories in India due to proximity to the location of the historical Sumatra earthquake soon after which yet another quake of magnitude 7.2 on the Richter scale had struck the territory that lies on the Burma microplate. The combined effect had devastated the coastal facilities in Andaman and Nicobar Islands. Major facilities such as air-force base, naval base, breakwaters, jetties, berths and other auxiliary facilities were damaged. The harbour area faced severe structural damage as can be seen in Fig. 5.2. Though the damage shown in the figures seems to suggest damage of an extreme nature, loss of human life due to sudden sweeping of the islands by the tsunami waves (Fig. 5.3) was much more monstrous. The direct and indirect measurements of run-up based on clear water markings can be seen from Fig. 5.4. In locations, where the levels are not available, sea levels at different time periods were used to obtain the relative levels and later absolute levels were obtained with reference to chart data. The vessels capsized in the dry dock. Figure 5.5 depicts the dense mangroves along the coast of Wandoor area. Even though many of these places were inundated to some extent, no loss of life was reported. In tribal areas, similarly, thick forest protected the coast and inundation was negligible. However, many low-lying islands in the Nicobar group were completely inundated and washed away. An interesting fallout of the earthquake is the dramatic change in the water levels in and around the Andaman and Nicobar Islands. Gahalaut and Catherine (2005) reported that the characteristics of the rupture that had occurred

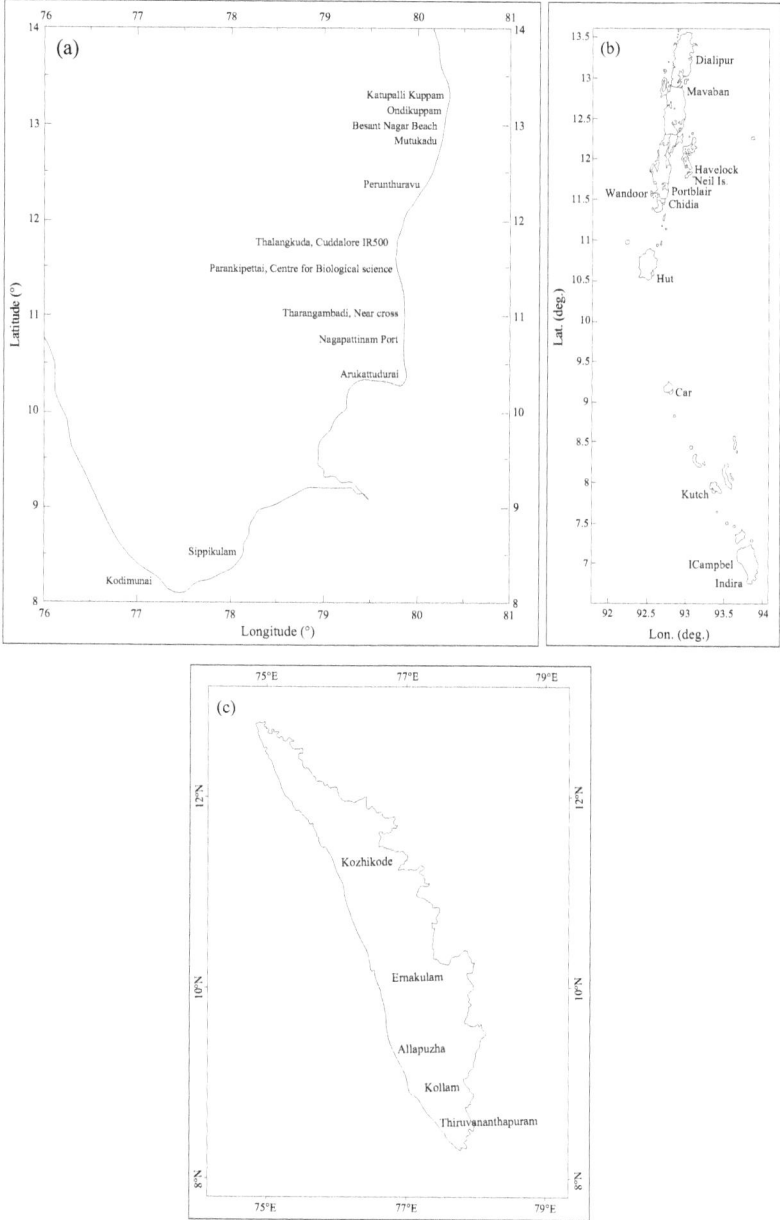

Fig. 5.1   Key map of the study area depicting (a) Tamil Nadu coast (b) Andaman and Nicobar islands and (c) Kerala.

Fig. 5.2   Extreme damage due to combined effect of the giant tsunami and earth-quake at Andaman Islands.

Fig. 5.3   Tsunami swept several locations in the Nicobar group of islands where land masses are narrow and low lying.

were quite different from south to north. Furthermore, the cracks that were formed in the plates at many places have become trenches. Hence, there appears to be a change in the general ground levels at many of the locations. One of the strongest pieces of evidence exhibiting this change is the difference between a photograph taken by the authors in May 2004 and a

(a)   (b)

Fig. 5.4 Direct and indirect run-up measurements. (a) Clear water mark at Chidiya Tapu; (b) Clear water mark at Bambooflat jetty.

Fig. 5.5 Mangroves in the Wandoor area effectively protected the habitat on the lee side.

photograph taken during the present survey at Bambooflat, which is shown in Fig. 5.6. In general, the entire group of islands seems to have tilted downwards in the south keeping North Andaman area as a pivot. Hence, the reduction in water levels is observed at Mayabunder and Diglipur which are the northernmost locations.

From the bathymetry around the Andaman and Nicobar Islands, it could be observed that a steeper shelf exists which is common around any island. This has resulted in a sudden amplification of the tsunami

(a)   (b)

Fig. 5.6   Direct evidence of sunk islands: (a) Picture of Bambooflat wharf on 10
May 2004 (during a tide of 1.58 m) and (b) on 21 January 2005 (during a tide of
1.33 m).

Table 5.1.   Survey stations and their run-up along the Andaman Islands.

Location	Latitude (N)	Longitude (E)	Run-up (m)
Chatham wharf	11°41′96″	92°43′55″	3.4
Haddo Container yard	11°41′04″	92°43′18″	3.1
Dry Dock I	11°40′31″	92°44′09″	3.3
Junglighat	11°39′33″	92°43′49″	3.8
Bambooflat Jetty	11°40′11″	92°42′34″	4
Hut Bay	10°34′41″	92°34′12″	6.1
Chidiya Tapu	11°28′26″	92°43′48″	3.9
Wandoor	11°35′14″	92°36′57″	2.8
Neil Island	11°50′21″	93°01′57″	2.1
Havelock Island	12°02′31″	92°58′54″	3.0
Mayabunder	12°53′96″	92°55′34″	2.9
Diglipur	13°16′16″	93°04′84″	2.9

waves while they propagated over the shelf. It is to be noted that the
fault line lies just around the 1,000 m contour on the western side. Hence,
the waves immediately after generation would have struck the islands. The
estimation of signatures for Little Andaman island and Nicobar group of
islands were based on indirect evidence such as debris trapped on the
branches of tress, videos and interviews with the locals and government
officials. The details of the survey stations and the run-up levels are listed
in Table 5.1.

The estimated run-up levels for the Andaman and Nicobar Islands are
given in Fig. 5.7. The highest run-up in Andaman Island seems to have

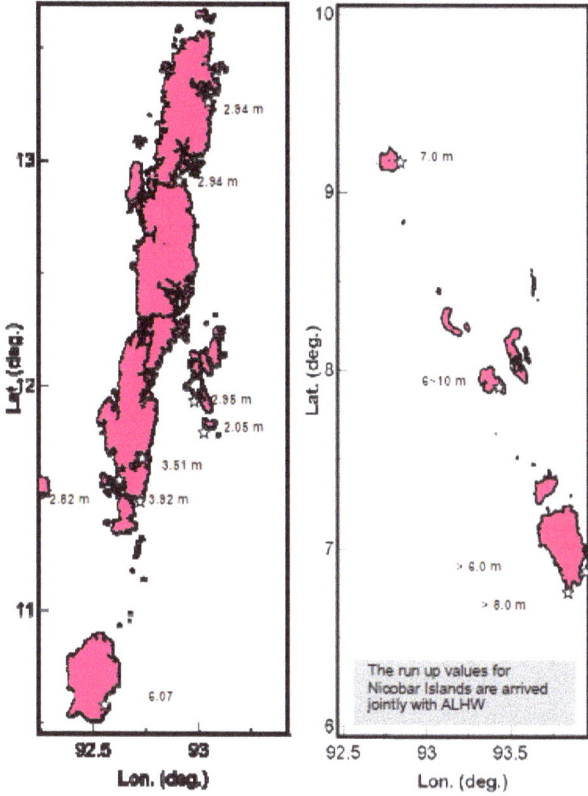

Fig. 5.7   Estimated run-up levels for Andaman and Nicobar Islands.

occurred at Chidiya Tapu, which is the southernmost point almost on the eastern coastline of the island. At the same time, on the western side, at Wandoor, the estimated run-up is about three quarters of the run-up observed at Chidiya Tapu. This difference seems to have been caused by the presence of Rutland just below Andaman Island. Moving northwards, at Port Blair's harbour area, an average run-up height of 3.51 m was observed. However, due to complex configuration of the port area, various installations like berths experienced different run-up levels (Fig. 5.8). The well-protected jetty at Junglighat faced severe damage due to the amplification of the waves inside the basin. Along this jetty and other jetties, many of the berthed vessels were lifted by the power of the tsunami and were grounded. Many of the vessels away from the jetties escaped the wrath of the tsunami. As the tsunami entered the harbour area, it would have directly hit the

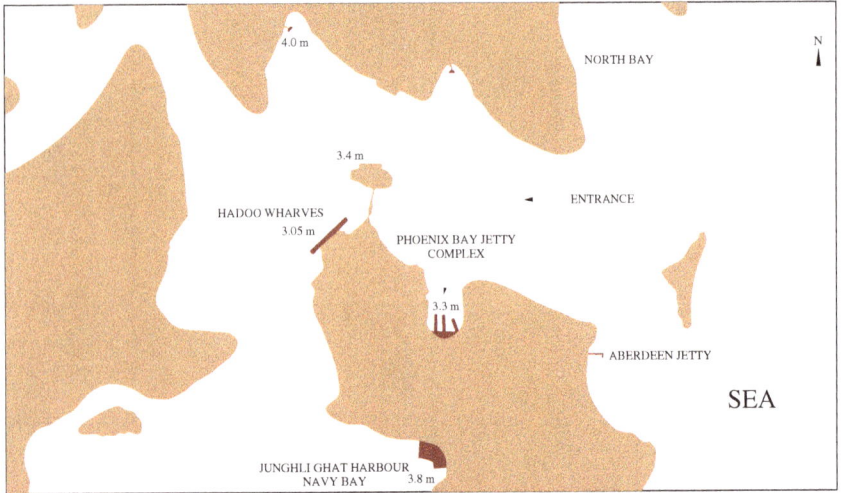

Fig. 5.8   Estimated run-up levels for Port Blair harbour.

Bambooflat bay resulting in the maximum run-up. However, the Chatham wharf and the Dry Dock-I (Phoenix bay complex) did not experience high run-up as they are located along the propagating direction of the tsunami. The Haddo wharf seems to have been protected by the presence of the Chatham Island.

Moving northeast to the islands of Havelock and Neil, as these small islands are located in the open waters, no dramatic amplification of the run-up seemed to have taken place. No significant structural damage was reported in this area. In the north, at Mayabunder and Diglipur, the run-up levels were up to about 2.9 m. Again, no significant damage to structures had taken place. In the south, the run-up levels increased from Hutbay to the southernmost tip of Indira Point. The run-up levels in general exceeded 6 m. Around the island of Katchal, the run-up level in general was close to 10 m. In the Great Nicobar Island, the run-up was close to 10 m. It is believed that in many places the run-up exceeded this magnitude. In the Nicobar group of islands, the tsunami and the quake seem to have coincided to produce the worst possible effect of sudden loss of habitat and property. In low-lying areas, for instance, near Wandoor, the seawater inundation of up to 2 km was reported due to a close-knit creek system which had resulted in marginal salinity intrusion into groundwater. However, at locations like Chidiya Tapu, Junglighat and Havelock, mangroves have effectively protected the habitats on their lee side.

5.2.3   Tamil Nadu coast

5.2.3.1   North Tamil Nadu coast (Phase 1)

In the first phase of the survey, a total of 38 locations were surveyed, and their geographical locations are provided in Table 5.2. Katupalliku-pam (northernmost part) was the survey starting point. In the absence of initial withdrawal, the tsunami reached at 8.45 am. A total of three waves were seen with the last wave at 10.45 am. At Kasimedu in Chennai (north of Chennai port), 20 dwelling units were completely destroyed. However, the rubble mound groynes protected the area from further damage.

Table 5.2.   Survey stations along north Tamil Nadu coast.

No.	Name of the place	Latitude (N)	Longitude (E)	Inundation (m)	Run-up (m)	Time of struck
1	Katupalli Kupam	13.18′24.4″	80.20′48.7″	300–500		A — 8.45
						C — 10.45
2	Ennore Creek	13.13′56.9″	80.19′51.7″	500		
3	Netukuppam	13.13′54.4″	80.19′47.8″			A — 8.50
4	Chinna Kuppam	13.12′35.3″	80.19′20.2″	500	3.08	
5	Ondikuppam	13.09′14.7″	80.18′17.4″		3.81	A* — 9.00
6	Marina beach, Chennai	13.02′24.9″	80.16′47.5″		2.23	A — 9.00
7	Light House, Chennai	13.02′04.9″	80.16′43.1″	700	2.27	A — 8.55
8	Fore Shore Estate	13.02′04.9″	80.16′35.3″	400	4.51	A — 9.00
9	Besant Nagar Beach	12.59′47.0″	80.16′14.2″	200	2.76	A — 9.00 C*
10	Thiruvanmiyur	12.58′36.4″	80.16′0.7″	100	3.65	
11	Kottivakkam	12.59′51.6″	80.15′48.1″	300	4.85	A — 9.00
12	Injambakkam	12.54′40.4″	80.15′19.0″		3.2	
13	Muttukadu	12.49′53.8″	80.14′51.0″		4.1	A — 9.00 C*
14	Kovalam	12.47′22.6″	80.15′10.1″		5.71	
15	Venpurusham, Mahabalipuram	12.35′53.5″	80.11′26.8″	500	5.74	
16	Periyakuppam	12.26′42.8″	80.8′35.3″	700	5.67	A — 10.30 C*
17	Perunthuravu	12.22′38.9″	80.05′31.3″		4.59	
18	Alamparai	12.16′16.2″	80.01′00.8″		4.75	
19	Kizhapettai Kuppam	12.08′00.4″	79.55′44.9″		3.29	A — 8.45 B* D — 11.30
20	Periya Kalapet Kuppam	12.01′51.6″	79.52′05.2″	300–500	5.79/6.93	A* — 8.45 D — 9.30

(*Continued*)

Table 5.2.   (*Continued*)

No.	Name of the place	Latitude (N)	Longitude (E)	Inundation (m)	Run-up (m)	Time of struck
21	Thazhanguda, Cuddalore	11.46′00.1″	79.47′34.3″	IR-500 m, 1500	3.66/9.08	A — 8.45
22	Silver Beach, Thevanampatti-nam, Cuddalore	11.44′22.6″	79.47′09.1″	1000	7.55/11.25	
23	Tamil Nadu Maritime Board, Port office, Cuddalore, Tamil Nadu Petrochem Ltd.	11.42′27.8″	79.46′30.0″	142	2.76	A — 8.45 B — 9.00
24	Annankoil	11.30′15.1″	79.46′13.5″	1500	11.12	
25	Parangipettai, Centre for Biological Sciences	11.29′27.7″	79.45′55.6″		2.32	A — 9.00 B* C
26	Thirumullaivasal	11.14′31.0″	79.50′37.4″	700	5.22	A* — 9.00
27	Poompuhar	11.08′36.1″	79.51′24.5″	100	5.72	
28	Tharangambadi Village	11.01′48.1″	79.51′21.1″		4.61	B* — 9.00
29	Tharangambadi, Near cross	11.01′32.3″	79.51′22.8″		8.56	A — 8.50
30	Karaikal Beach	10.54′49.1″	79.51′05.01″	IR 500 m, 1000	4.48	9.00
31	Nagore	10.49′21.4″	79.51′03.6″	1000	5.81/6.5	A — 9.10 B* — 9.30
32	Aryanatu Theru, Nagapattinam	10.46′08.8″	79.51′00.5″	IR 500 m, 1000	4.68	
33	Nagapattinam Port	10.45′47.4″	79.50′57.4″	300	4.59	A — 9.30 B* — 9.35 R 500 m
34	Akkaraipettai	10.44′34.5″	79.50′58.2″	IR 100 m,	4.02	A — 9.15
35	Kallar Village	10.44′21.1″	79.50′57.2″	1500	4.3	A — 9.15
36	Velankanni Church	10.40′48.4″	79.51′10.0″	500	4.9	
37	Arukattu Thurai, 10 km from Point Calimere	10.23′38.4″	79.52′01.3″		6	A — 8.30
38	Athirampattinam	10.19′18.3″	79.23′49.7″	300	4.48	

Note: IR: Initial receding, R: Receding, A: First wave, B: Second wave, C: Third wave.
*The most destructive wave.

The tsunami arrived the Marina beach south of Chennai Harbour along the southeast coast of India at 9 am. The cars from the parking lot were lifted up and washed out to the main road by about 100–200 m. The waves struck three times with the first one being the biggest. In Thiruvanmiyur (south Chennai), the wave surge height was up to 5 m and the third wave was the strongest. The waves damaged in total a 1.5 m tall reinforced wall of 0.3 m thickness. No significant initial rundown was observed. Initial recession was only 10 m at Injambakkam. The third wave was reported to be the largest at 9 am. In Muttukadu, considerable destruction was observed. Figure 5.9 shows the damaged scene of a newly constructed house. Evidence of boats washed far into the land and the village up to a distance of about 300 m was completely deserted. The height of the wave run-up was about 6 m. The water receded up to 1 km offshore at Kovalam before the first wave advanced the shoreline. The initial recession was approximately 1 km at Venpurusham near Mahabalipuram.

The seawall in Pondicherry or Puducherry town reduced the impact of the tsunami wave. In Periya Kalapet Kuppam south of Pondicherry, accretion of sand was found after the tsunami impact. Thalanguda in Cuddalore district observed an initial withdrawal of 500 m before the first wave surpassed the shore. The numerous coconut trees along the shore region had fallen sequentially due to the erosion of sand during the return flow and

Fig. 5.9   Structure failure due to scouring of foundation at Muttukadu.

the sand deposition was seen at south in the Pennaiyar river mouth. In silver beach of Thevanampatnam, Cuddalore district, the first wave struck at 8.45 am with an inundation of up to 1 km. The river mouth was found to be covered in heaps of sand. There was a surge in the sea level of about 3–4.5 m initially. The stagnation time was for about 5 h at a height of 1.3 m above the ground level. A 2-tonne roller was carried from the shore to a distance of about 200 m. A stagnation water mark was observed at 3.2 m from mean sea level (MSL).

Annan Kovil observed two waves with the first wave at 7–8 m of height at 8.45 am and a higher second wave of about 10 m in height. A lamppost was bent in the southwest direction, indicating the wave direction here. The inundation was about 1.5 km inland (Fig. 5.10). In Thirumullaivasal, the initial arrival time of the wave was 8.50 am with an inundation of 700 m inland. In 1950, the Dutch had built a weir which had been overlaid by a road. The tsunami completely washed out the road, exposing the weir.

The approach road of Karaikal beach was severely damaged and the river mouth was exposed completely (Fig. 5.11). In Nagore, an initial withdrawal of 500 m was observed in Aryanatu Theru, Nagapattinam district at 9.10 am where the old miniport was damaged. The coconut trees were washed away and newly formed sand deposition was found downstream south. The

Fig. 5.10   Picture taken from light house, Annankovil.

Fig. 5.11   River mouth opened at Karaikal.

Nagapattinam port, one of the worst affected areas with just 0.6 m above MSL, received the wrath of the tsunami, leading to heavy causalities as the thickly populated town is adjacent to the coast. The breakwater of the Nagapattinam port was partially washed away and boats were severely damaged by collision experienced due to the penetration of long waves (Fig. 5.12). A sand dune near the shore was washed away completely at Velankanni and sand deposition was found downstream (south).

The village Athiramapatnam, was the final survey point of the first phase. The damage was only to boats and nets and an initial warning was given to the local people at 8.30 am by the officials about the strange behaviour of the sea. Aliyathi plants in the seashore reduced the damage (Fig. 5.13). Table 5.2 gives the survey data of Phase 1.

5.2.3.2   *South Tamil Nadu coast (Phase II)*

Based on the field data, it was observed that the tsunami waves, which struck the southern Tamil Nadu coast, were the diffracted waves from the Sri Lankan island. Further, it was observed that along the southeast coast, the second tsunami wave created a larger destruction than the first one. The first one was just a bore (piling of water like a wall) and hence resulted in rise of water level and gave a warning of a tsunami. Also, a receding

Fig. 5.12   Collapsed boats at Nagapattinam.

Fig. 5.13   Vegetation along the coast.

of the shoreline ranging from 300 m to 1 km was reported in the southeast coast. Table 5.3 lists the survey stations and the tsunami characteristics.

When compared to the southeast coast of Tamil Nadu, the southwest stretches felt the maximum impact of the tsunami and huge losses of both human lives and property were reported. The reason may be the protection

Table 5.3.   Survey stations, run-up and inundation level along southern coast.

Name of place	Latitude (N)	Longitude (E)	Inundation (m)	Run-up (m)
Sippikulam	8°31′4.5″	78°07′22.3″	25	—
Threspuram	8°48′51.4″	78°09′47.8″	50	—
Veerapandipattinam	8°31′4.5″	78°15′9.7″	220	2.34
Alanthalai	8°27′4.2″	78°05′53.7″	40	4.07
Periyathalai	data not collected	data not collected	0	—
Overy	8°16′41.3″	77°53′31.3″	60	3.16
Idinthakarai	8°10′29″	77°44′20.4″	190	—
Kootapuli	8°08′41.2″	77°38′25.2″	140	5.68
Arokiapuram	8°07′9.6″	77°33′31.7″	400	4.27
Alangramatha street	8°05′3.0″	77°33′8.2″	60	4.295
Kovalam	8°04′54.7″	77°31′28.5″	90	—
Kezhamanakudy Thurai	8°05′18.5″	77°29′16.8″	90	6.95
Melamanakudi Thurai	8°05′20.9″	77°28′50.9″	600	—
Sothavilai Beach	8°05′50.5	77°27′01″	350	—
Pallanthurai	8°5′57.2″	77°25′51.1″	40	6.37
Pillaithoppu Thurai	8°07′29.8″	77°20′24.5″	520	4.8
Colachel	8°10′20.3″	77°15′31.6″	310	4.35
Kodimunai	8°10′39.7″	77°14′20.7″	160	—
Keezh Madaalam	8°12′10.7″	77°12′53.1″	200	4.065
Enayam	8°13′9.6″	77°11′5.5″	90	3.01
Thenkaipattinam	8°14′20.1″	77°10′7.7″	90	1.11

on the eastern side created by the shadow of the Sri Lankan island and the diffracted waves directly hit the southwestern coast.

Among the tsunami-affected areas in the southwest Kanyakumari coast, Colachel, Pilla Thoputhurai, Sothavilai beach, Melamanakudi Thurai and Kezhamanakudy Thurai were the worst-affected villages where a death toll of above 700 was reported. From the field investigation, it was found that the beach slope in Colachel area was very mild and the beach was entirely open to the sea without any coastal protective structures. Further, it was observed that the terrain was flat in nature and hence the waves inundated up to a distance of 500 m–1 km from the shoreline. Also, it was reported that the canal flowing along the coast silted up with debris and beach sand. Interestingly, the jetty at Colachel harbour did not suffer damage.

The fury of the tsunami waves that struck the southwest coast can be better understood from the Fig. 5.14, which shows the remains of a bridge at Melamanakudi Thurai. As the tsunami waves struck laterally, the mid spans of the bridge constructed at the river mouth were carried to the river bed and the end spans were thrown along the riverbanks.

Fig. 5.14   Remains of a bridge at Melamanakudi Thurai.

Fig. 5.15   Damaged seawall and church at Kezhamanakudy Thurai.

Figure 5.15 shows the extent of damage caused to the seawall and nearby church at Kezhamanakudy Thurai and it was reported that the displacement of armour stones during the tsunami caused severe threat to human lives and the residential structures.

Huge loss of human life and property were reported from Pillathoppu Thorai and Sothavalam beach, as these places were open to sea without any coastal protection. Further, these areas are of flat terrain and hence the waves surged up the land and inundated up to a distance of above 500 m.

The stretches of Thengapattanam, Keezhmidalam, Kovalam, Algramatha Street (Kanyakumari town) and Arokiapuram are protected either by seawalls or groyne field. In the above stretches, the loss of property was reported to be less. Further, the beach slope was found to be relatively steeper, the land was well elevated above the MSL, which reduced the inundation level and hence the damage. Severe damage to fishing boats and catamarans and some damage to nearby houses but no loss of human life were reported from these areas.

The coastal stretches of Tirunelveli district were less affected by the tsunami waves when compared to the neighbouring Kanyakumari district. The effect of tsunami was felt at some pockets of Kootapuli, Idinthakarai and Uvari, where the average land level was nearer to the MSL. Along the coast, water receding after the first wave has been observed. Figure 5.16 shows the sequential attack of tsunami on a groyne that was under construction at Periyathalai. It was reported that the maximum run-up reached a height of 5.5 m from MSL, which submerged the groyne and can be clearly seen in Fig. 5.16.

In Veerapandipattinam, as the beach slope is mild, the tsunami waves had surged up to a distance of 220 m causing destruction to residential buildings. Except Veerapandipattinam, the other stretches of Tuticorin district were least affected, and no major threat to both human lives and property was reported.

Tamil Nadu was one of the most affected coastal states of India during the 2004 tsunami. The measured run-up level along the Tamil Nadu coast is shown in Table 5.4. Estimation of run-up levels in these areas was based on the signatures of water levels on structures, numerous interviews with local public, Engineers of the Public Works Department on duty. The key plan of the study area is shown in Fig. 5.17. The run-up and the inundation

Fig. 5.16   The sequential attack of tsunami on the groyne at Periyathalai.

Table 5.4.   Measured Tsunami run-up height along Tamil Nadu coast.

Location	Tsunami run-up height above MSL
From north Tamil Nadu to south Chennai (up to Muttukadu)	2–5 m
Kovalam to Pondicherry	4–7 m
Cuddalore to Athirampattinam	5–11 m
From north of Tuticorin to Kanyakumari	2–5 m
West coast of Tamil Nadu	4–7 m

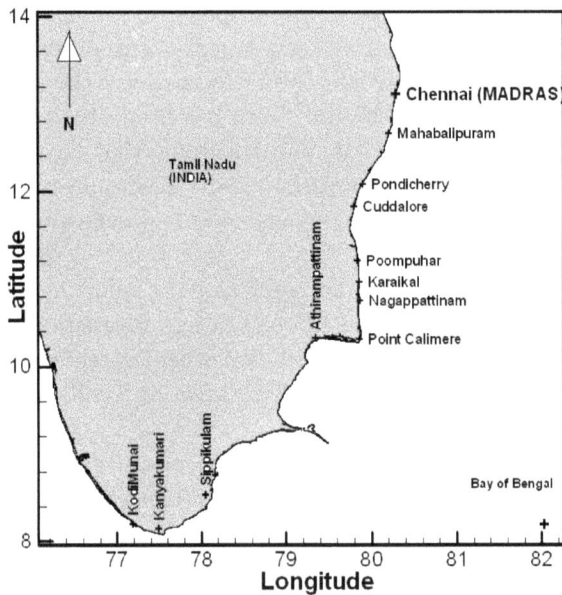

Fig. 5.17   Key map of the study area depicting Tamil Nadu coast.

distances along the northern Tamil Nadu coast are shown in Fig. 5.18, while that along the southeast and western coast are projected in Fig. 5.19. For further details refer to Sundar (2005) and Sundar *et al.* (2007).

5.2.4   *Kerala*

The impact of the tsunami on the Kerala coastline was comparatively less than that felt in Andaman and Nicobar Islands and Tamil Nadu. Tsunami waves of a considerable impact was felt across the southern coastal districts of Allapuzha, Eranakulam, Kollam, Kozhikode and Thiruvananthapuram.

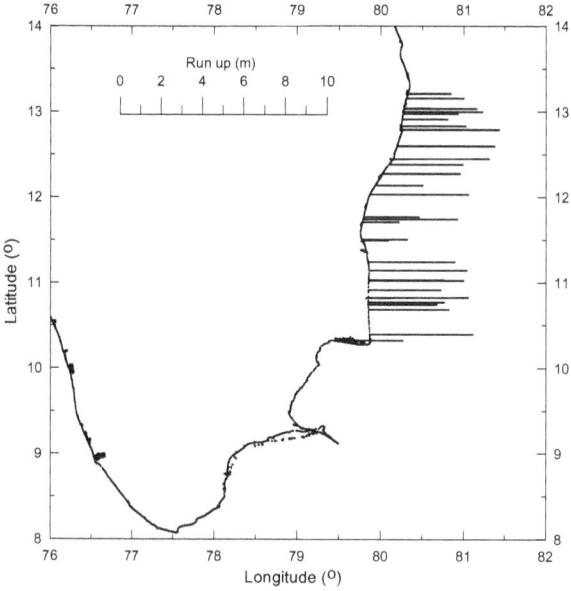

(a)

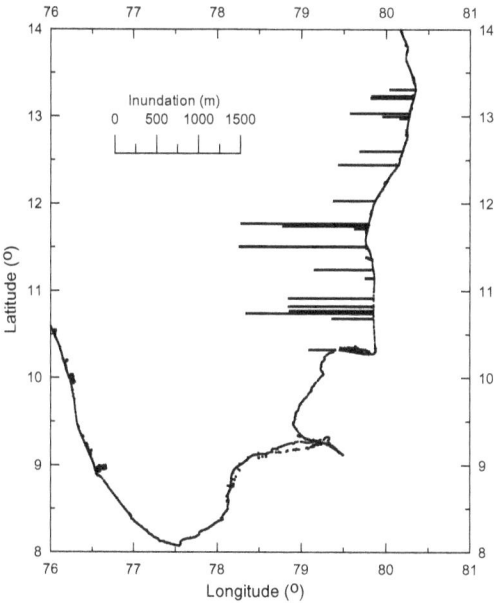

(b)

Fig. 5.18   (a) Measured run-up height along northern coast of Tamil Nadu.
(b) Measured inundation distance along northern coast of Tamil Nadu.

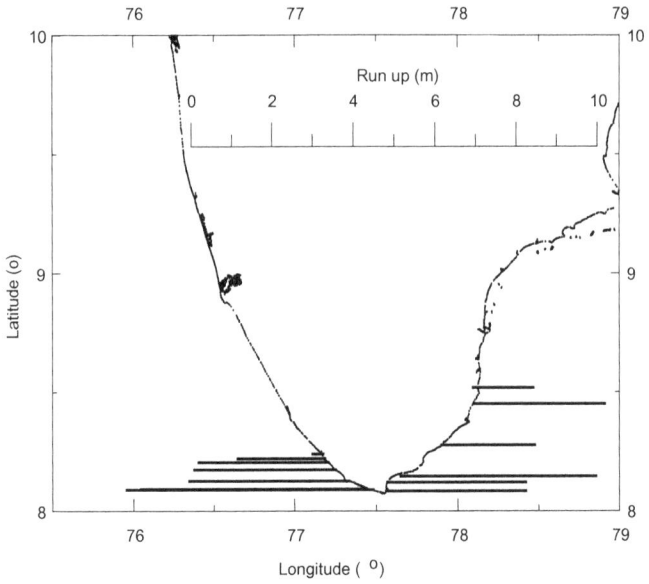

(a)

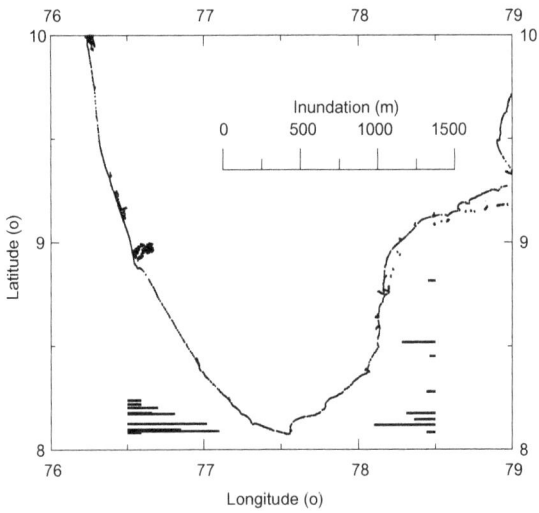

(b)

Fig. 5.19   (a) Measured run-up height along the southeast and western coast of
Tamil Nadu. (b) Measured inundation distance along the southeast and western
coast of Tamil Nadu.

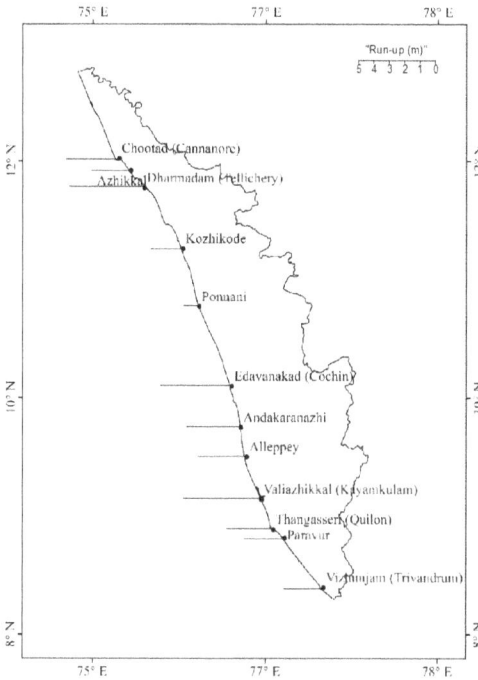

Fig. 5.20   Extent of run-up along the prominent location in Kerala coast during the 2004 tsunami.

The reasons for the subdued tsunami activity at several locations on the Kerala coast were the absence of resonance amplifications, and defocussing of tsunami energy (due to bathymetry), thereby creating zones of divergences where the tsunami amplitudes were smaller.

The tsunami after impinging the Sri Lankan island diffracted around the tip of the Indian peninsula and exhibited its devastating effects along the west coast, i.e., maritime state of Kerala. The run-up levels at a few selected locations along the Kerala coast are projected in Fig. 5.20 (Kurian et al., 2006).

5.3   Tidal Measurements

The tide gauges located in the major ports along the Indian coastline possess a continuous record of the tidal variations. The tidal record measured after the incidence of 2004 Indian Ocean tsunami at Kochi, Tuticorin, Chennai and Visakhapatnam ports (Sheth et al., 2006) in Figs. 5.21(a), 5.21(b), and 5.21(d), respectively. The time of landfall of the tsunami was marked.

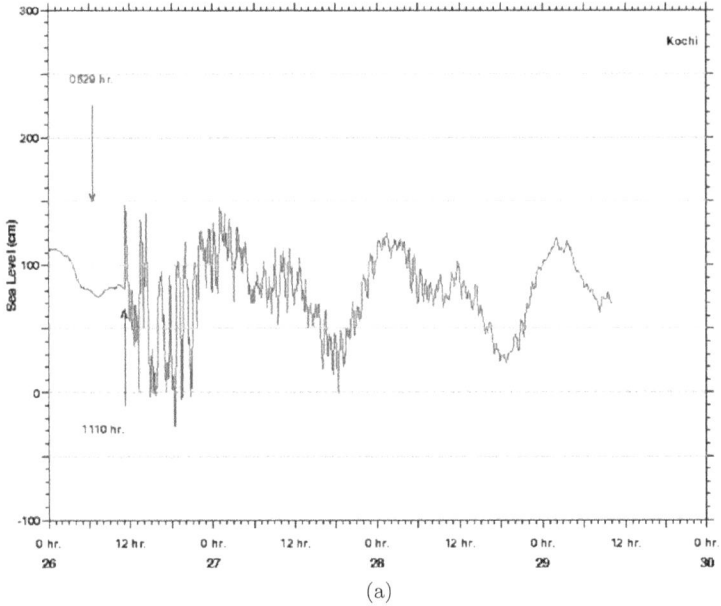

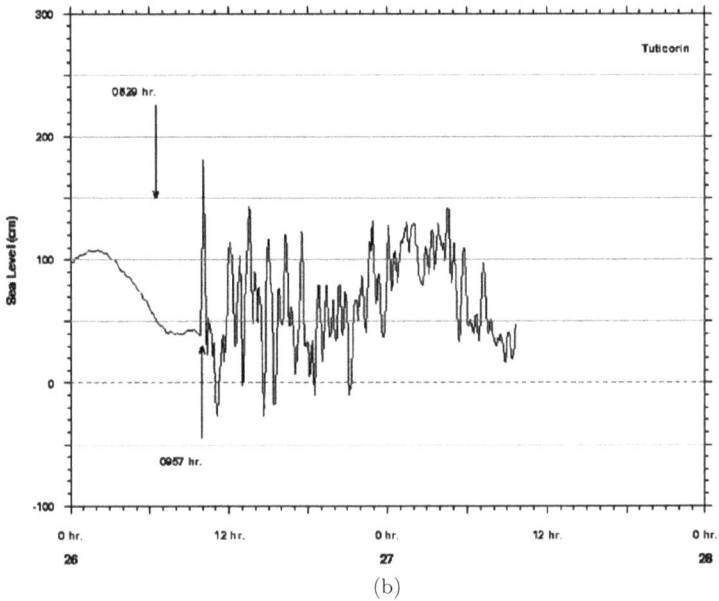

Fig. 5.21   (a) Measured variation of tide in Kochi port (— NIO, 2005).
(b) Measured variation of tide in Tuticorin port (— NIO, 2005). (c) Measured
variation of tide in Chennai port (— NIO, 2005). (d) Measured variation of tide
in Visakhapatnam port (— NIO, 2005).

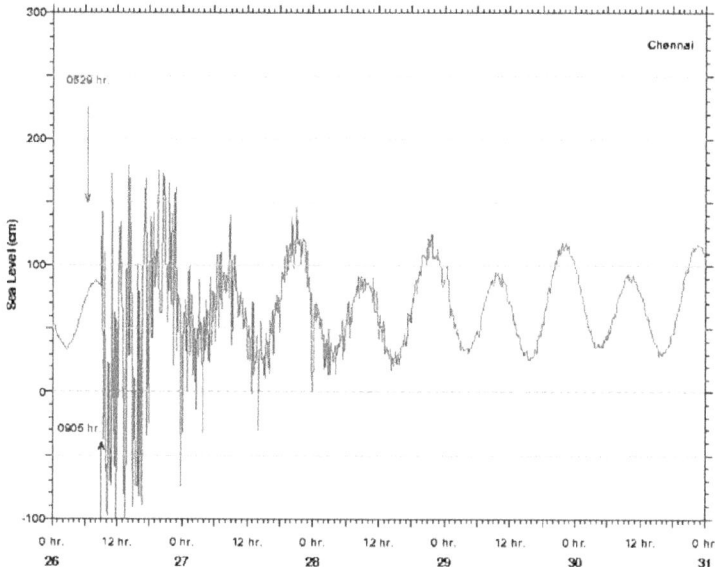

(c)

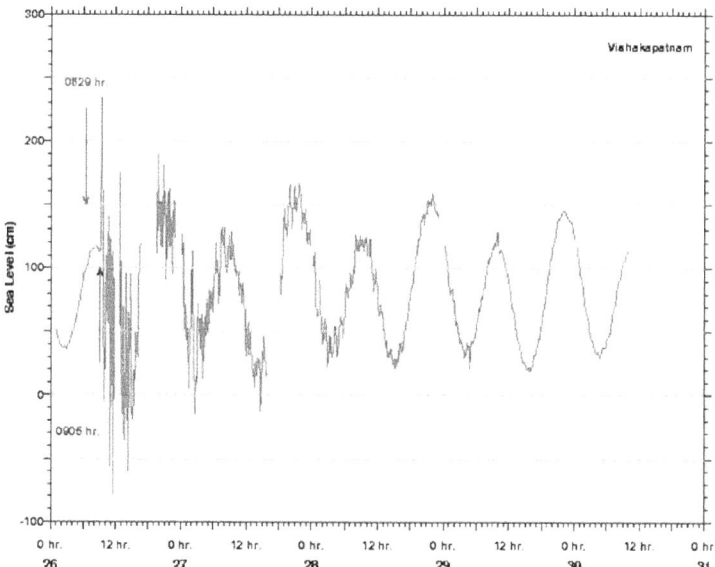

(d)

Fig. 5.21   (*Continued*)

It can be seen that the smooth variation in the water surface elevation was perturbed for the following couple of days. This meant that the waves of few minutes up to 20 min had superposed over the tidal wave. The resulting oscillation affected the regular loading and unloading operations, if the amplitude exceeded a threshold limit. More importantly, the manoeuvring of the vessels inside the harbour was affected.

Tide gauge measurements from south to north have shown that water levels have changed considerably at several locations. Nirupama *et al.* (2006) indicated that the changes in the water level could be attributed to the possible trapped modes in the shelf of Andaman and Nicobar Islands. However, tide gauge measurements after the great Indian Ocean tsunami indicate that there is a permanent change in the MSL (Suresh *et al.*, 2006). However, there is no appreciable change in water levels in Havelock and Neil islands. This appears to be due to the differential fracturing of the plates.

5.4   Summary

A post-tsunami field survey has been carried out to understand the tsunami wave propagation and its inundation levels along east coast of India. The survey results will be useful in predicting the cause of severity of tsunami wave attack and hence, will be helpful for disaster mitigation measures in the future.

The devastation of tsunami has been encountered in terms of floatation during water up rushing, inundation during the stagnation period if any, and under scouring during receding. The impact force can be estimated from the flood velocity on the land and the impact due to floating debris is severe. All the reinforced concrete construction with sufficiently deep foundation has survived and shallow foundation has been fully exposed.

References

Gahalaut, V. K., and Catherine, J. K. (2006). "Rupture characteristics of 28 March 2005 Sumatra earthquake from GPS measurements and its implication for tsunami generation." *Earth and Planetary Science Letters*, 249(1–2), 39–46.

Kurian, N. P., Baba, M., Rajith, K., Nirupama, N., and Murty, T. S. (2006). "Analysis of the Tsunami of December 26, 2004, on the Kerala Coast of India — Part I: Amplitudes." *Marine Geodesy*, 29(4), 265–270.

Nirupama, N., Murty, T. S., Nistor, I., and Rao, A. D. (2006). "Persistent high water levels around Andaman and Nicobar Islands following the 26 December 2004 tsunami." *Science of Tsunami Hazards*, 24(3), 183–193.

Sheth, A., Sanyal, S., Jaiswal, A., and Gandhi, P. (2006). "Effects of the December 2004 Indian Ocean Tsunami on the Indian Mainland." *Earthquake Spectra,* Vol. 22, No. S3, pages S435–S473.

Sundar, V. (2005). "Behaviour of shoreline between groin field and its effect on the tsunami propagation." Proceedings 5th International Symposium on Ocean Wave Measurement and Analysis, WAVES 2005, Madrid, Spain, 3–7 July, Paper No: 323.

Sundar, V., Sannasiraj, S. A., Murali, K., and Sundaravadivelu, R. (2007). "Run-up an inundation along the Indian Penninsula including Andaman Islands due to great Indian Ocean Tsunami." *Journal of Waterway, Port, Coastal, and Ocean Engineering, ASCE,* 133(6), 401–413.

Suresh, I., Neetu, S., Shankar, D., Shenoi, S. S. C., Shetye, S. R., Sundar, D., Shankar, R., and Nagarajan, B. (2006). "The 2004 Indian Ocean Tsunami: Description of the event and estimation of length of the tsunami source region based on data from Indian tide gauge." Proceedings of the Eleventh Asian Congress of Fluid Mechanics (11ACFM), 22–25 May.

Chapter 6

Post Facto Evaluation along Tsunami Affected Stretches

6.1 General

Planning of Coastal structures against extreme events (storms, tsunami, etc.) is necessary to reduce the impact of natural hazards on the coastal communities and infra structures. This chapter aims at the planning of coastal structures taking into account the effects of the aftermath of the tsunami, triggered due to the earthquake off Sumatra, of magnitude 9.1 on Richter scale, on 26th December, 2004. In India, the maritime states of Tamil Nadu situated on its south east coast and Kerala along the South west coast were badly affected. The different types of coastal protection structures considered includes seawalls, groynes, combination of groynes and seawalls, artificial beach nourishment, buffer blocks, plantations, etc. Apart from the structural measures to be discussed in this chapter, the roots and stems of plants are natural traps for sand particles that would otherwise be carried away by wind, currents and waves. A brief understanding on how the vegetation plays an important role in terms of tsunami resistance is reported in Chapter 8 of this book. A flat beach is more favourable for such plantation. In addition, marsh vegetation acts as a buffer against wave action and tsunami to an extent. Vegetation as a protection can reduce flow induced loads as well as increase soil strength. Roots can increase the strength by protecting the grains on a micro scale or by reinforcing them. Mangrove forests are the natural vegetation of many tropical coasts and tidal inlets; they form a highly productive ecosystem, a nursery for many marine species Schiereck (2001).

The protection measures are selected based on the factors such as littoral drift, beach characteristics, and coastal morphological features like estuaries, bays and tidal flats. The planning for coastal protection structures against tsunami waves is economically not viable. Hence, the structures proposed here are intended to gain beach area, which will act as a buffer zone facing the brunt of the tsunami waves and thus minimizing the inundation distance of sea water. The protection measures suggested only for typical locations are herein reported.

6.2 Background of South India

Tamil Nadu is located in the South East side of the Indian Peninsula, its coastline stretches for about 1050 km bordering the Bay of Bengal. The coastline extends from Pulicat (N 13°26′38.97″ E 80°19′34.61″) in the north to Erayamanthurai (N 8°17′41.51″ E 77°5′37.52″) in south-west at Kanyakumari district. This shoreline stretch consists of many estuaries of ecological importance, Major and Minor ports, Fishing harbours, monuments of international heritage, tourist locations, pilgrimage centres, etc. Kerala Coast dynamic in nature situated along the south west of Indian peninsula is of length of about 590 km, facing the Arabian Sea extending from Latitude 8°15′ N to 12°85′ N and longitude 74°55′ E to 77°05′ E.

6.3 Coastal Protection Measures/Erosion Mitigation Measures

6.3.1 *Tamil Nadu*

The stretches of the coast exposed to the tsunami disaster needs to be addressed with a critical analysis of the local geo-morphology, wave climate, cyclone prone nature of the coast etc. In addition to restoring the coasts from damage due to tsunami, it should protect the shore from further damages due to combined wind-wave action. A few prominent stretches along the Tamil Nadu coast (as shown in Fig. 6.1) are discussed in the following sections. It is to be noted that groyne fields at Royapuram and Kanyakumari were implemented along the Tamil Nadu coast a few months prior to the Indian Ocean tsunami of December 2004. These measures in addition to providing coastal protection, it was instrumental in mitigating the impact of tsunami waves. In the event of cyclone, these structures act as buffers to wave forces acting on the coasts. The behaviour of shoreline between the groynes protecting the Royapuram coast in the event of Tsunami is presented and discussed in detail in Chapter 4.

Ennore Creek (N 13°13′56.9″; E 80°19′51.7″)

The river Kosasthalayar discharges into the Bay of Bengal at Ennore. The Ennore port is located on the north of this river mouth. The breakwaters

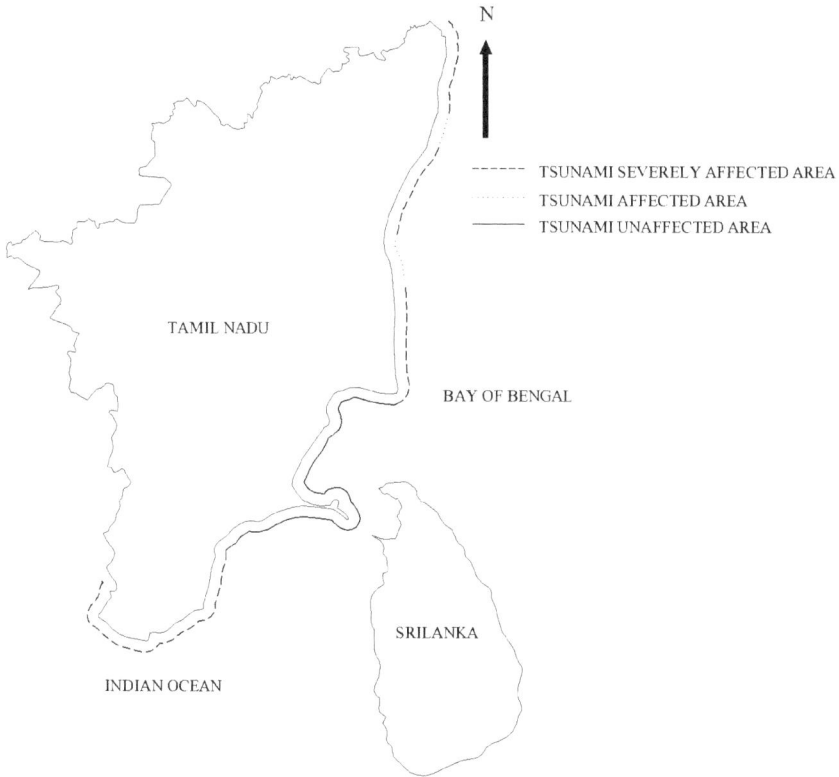

Fig. 6.1   Tamil Nadu location map (to be redrawn).

of Ennore port are acting as permanent littoral barriers and hence, trapping the sediments in to the river mouth. These trapped sediments have choked the river mouth. However, during the tsunami, the mouth of this river was opened due to the penetration of water mass from the ocean. A proposed solution was to construct groyne fields in two different phases each of which consisting of about 9 number of groynes. A schematic representation of the stretch of the coast from Eranavoor to Ennore port along with the completed groyne field for stretch I is depicted in Fig. 6.2.

The implementation of the said coastal protection measure is shown in Fig. 6.3. The mouth of the creek remain open as well as being effective in protecting the stretch of the coast.

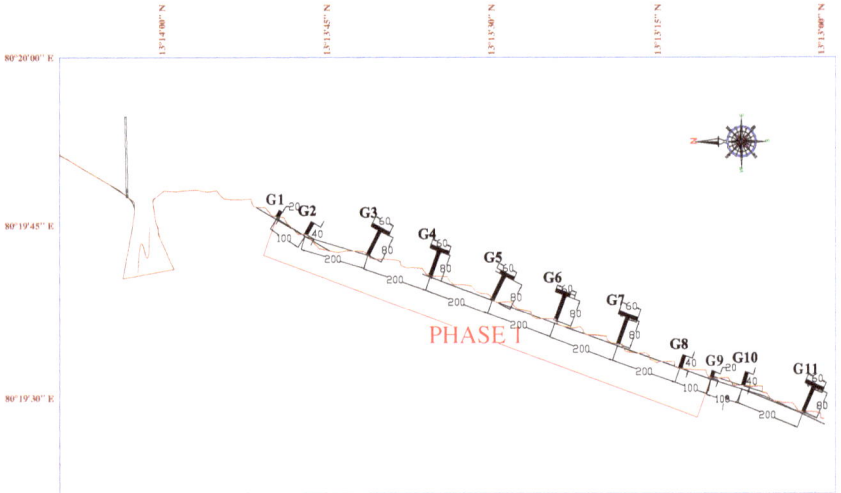

Fig. 6.2   Schematic representation of the stretch of the coast between Ernavoor and Ennore creek.

Fig. 6.3   Images of the shoreline evolution post implementation of phase I.

Near Cuddalore Port (N 11°42′25.5″; E 79°46′33.34″)

Singarathoppu is an inhabited coastal village situated on the north of the northern training wall leading to the Cudallore port. It is bounded by Bay

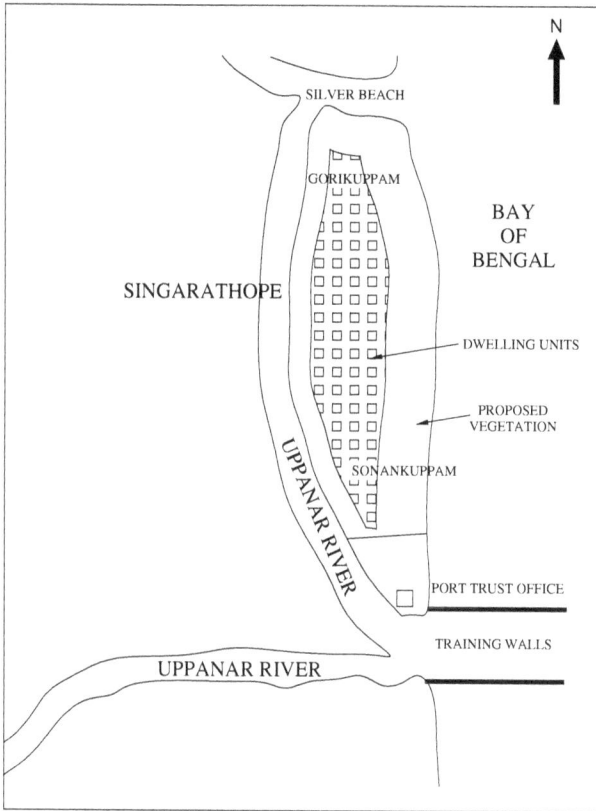

Fig. 6.4   Location map of Singarathoppu village.

of Bengal on its east and by Uppanar river on its west as shown pictorially in Fig. 6.4. As the dwelling units over this area is more towards the river Uppanar, the tsunami from the Sea did not affect the village significantly. However, it suffered maximum damage from the ingress of water due to tsunami penetrating into the river and flooding through its banks, because of which the fishing boat landing facilities were severely affected resulting in huge damage to the coastal community. The proposed safety measure from such events is to shift the dwelling units away from the river bank and implementation of dense vegetation in between the river bank and the dwelling units. The open land area facilitated easy inundation of sea water during tsunami and hence buffers zones using vegetation have been suggested. The roots of the vegetation cover can strengthen the soil and

Fig. 6.5   Variations at Singarathope.

reduce the risk or erosion, in addition to serve as a mitigation measure against run-up and inundation. The changes observed in the location prior to the 2004 tsunami and post implementation of vegetation is shown in Fig. 6.5.

Threspuram (N 8°48′55.4″; E 78°09′47.6″)

Threspuram is a small fishing village located north of the Tuticorin harbour. Owing to the predominant net northerly drift this stretch of this had been experiencing significant erosion. A proposal was made to construct a pair of groynes, the northern groyne with a long extending arm providing shelters from the wave action in the most predominant direction and the southern groyne was positioned adjacent to a drain mouth, the layout of which is shown in Fig. 6.6(a). During the construction phase additional three numbers of groynes were planned and constructed so as to facilitate larger boat landing area in its vicinity in relatively calm waters. Views of the coast of Threspuram during 2006 and 2017 are projected in Fig. 6.6(b). This measure will serve its purpose of sheltering the fishing boats and in the event of extreme coastal hazard like a tsunami it can act as buffers in reducing the inundation.

Idinthakarai (N 8°10′40.5″; E 77°44′46.8″)

This is a fishing village located in Thirunelveli district, with persistent erosion problem. The coast is mostly oriented along the east-west direction and

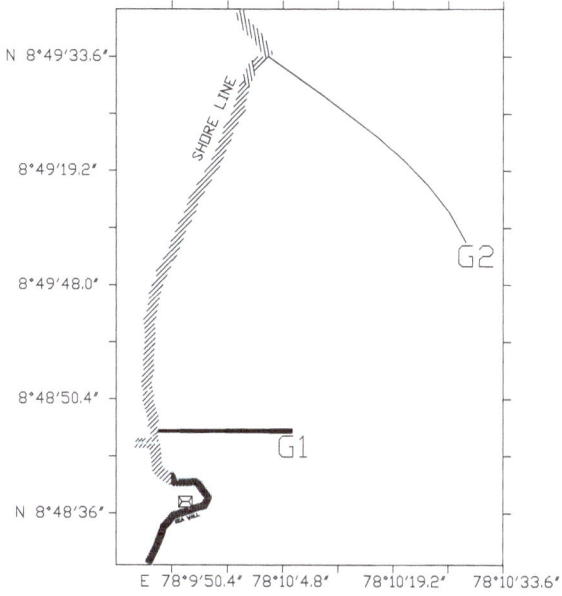

Fig. 6.6(a)   Therespuram layout.

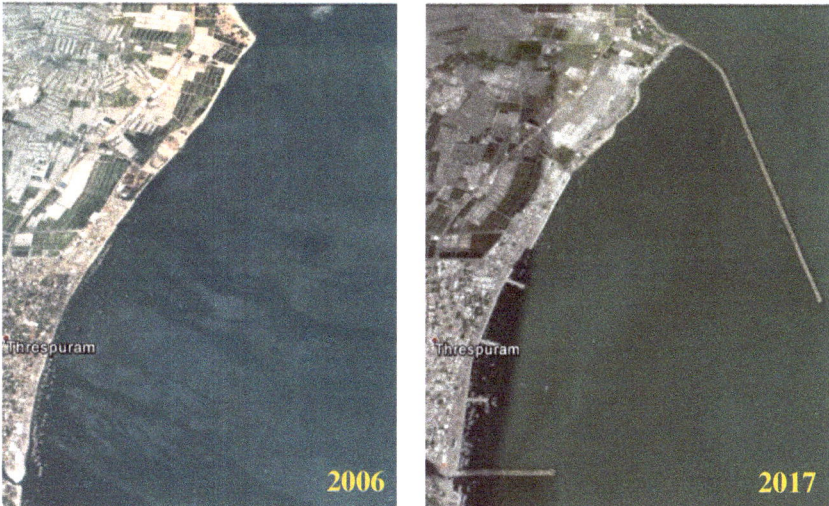

Fig. 6.6(b)   Shoreline variations at Therespuram.

the prominent net sediment transport is along the north-eastern direction. Four numbers of groynes of varying lengths was proposed to be constructed along a coastal stretch for about 1 km as shown in Fig. 6.7(a). The shoreline changes at Idinthakarai projected in Fig. 6.7(b) clearly demonstrates the effectiveness of protecting against erosion as well as winning the lost beach. Furthermore, the local fishing community is happy with their livelihood being enhanced as the beaches formed in between the groynes are serving as their boat landing facility. The beaches thus formed are likely to act as buffers during a possible tsunami.

Kanyakumari (N 8° 05′ 19.0″; E 77° 33′ 22.0″)

Kanyakumari, is a coastal district located in the southernmost tip of the Indian sub-continent. Several developmental activities take place and the livelihood of majority of the population is dependent on fishing related activities. The densely populated and highly exposed coastal stretches demands vigilantly planned protection measures that would mitigate the effect/impact on the coasts. In the early mid-seventies, a groin field comprising of 7 number of groynes, with the longest southernmost groyne was constructed along south-eastern coastal frontier. This groyne field was not

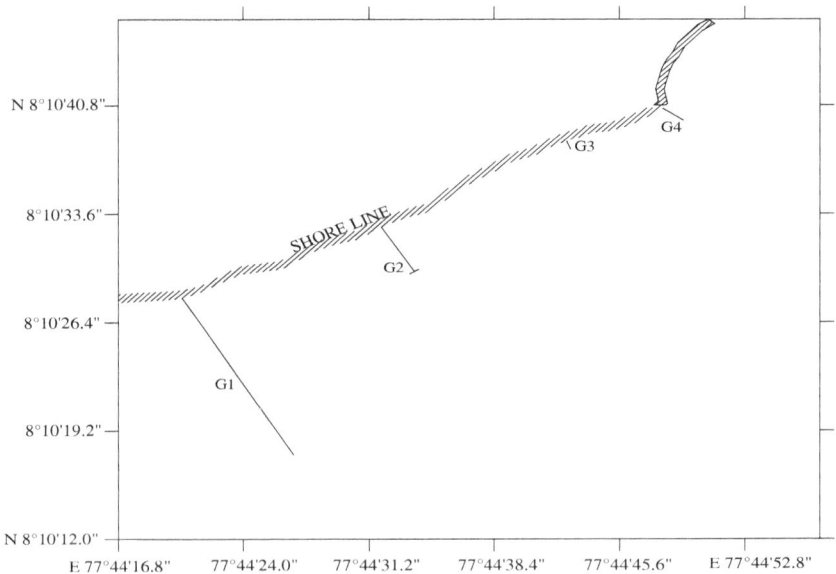

Fig. 6.7(a)   Layout for coastal protection at Idinthakarai.

Fig. 6.7(b)   Beach advancement near groyne field at Idinthakarai.

serving its purpose as the coast is characterised with rocky outcrops with less longshore sediment transport. The fishing community used an existing fish landing jetty along a neighbouring for parking their crafts that lead to a strong revolt from the locals leading to even loss of life and continuous unrest among the communities of the two villages. This being the background, the high energy tsunami waves diffracted on reaching Srilanka, and penetrated into the sheltered zone of the Kanyakumari coast. The diffracted tsunami waves oriented and propagated predominantly in the northerly direction i.e. almost perpendicular to the coast. The western stretch of Kanyakumari coast were characterised by numerous outcrops that aided in energy dissipation of tsunami waves, whereas, the eastern stretch experienced damage. Therefore, a proposal was made to extend only the length of southernmost groyne, G7 to protect the coastal stretch from further potential extreme waves/events. As a measure to reduce the inundation of tsunami in future and at the same serving as a meaningful investment, the southernmost G7 groyne was extended to conserve the onsluaght of waves along the coast between G6 and G7 groynes. The layout of the proposed groyne field is shown in Fig. 6.8(a). The shoreline changes over the years after the extension of G7 is shown in Fig. 6.8(b), and it can be evidently noted that the beach area is stabilized with the aid of these groynes.

Kodimunai (N 8° 11′ 04.27″; E 77° 13′ 54.93″)

This study area lies along the west coast of Tamil Nadu, where, accelerated rate of erosion had been witnessed due to the violent action of waves approaching the coast from the Arabian Sea, particularly during South-West monsoon i.e. between May and September months every year. The images of the location prior to implementation of protection scheme is shown in Fig. 6.9(a). A field survey/study was carried out over a stretch of about 3 km along the shore and about 1 km into offshore (up to a 10 m water depth). This stretch of the shallow waters is characterised by parallel seabed contours with several rock outcrops. A comprehensive analyses of the wave data have been carried out in order to determine the wave characteristics for the numerical model study. Based on a detailed study, six groynes were proposed for this area, of which two groynes at Kurumbanai, one groyne at Vaniyakudi, two groynes at Kodimunai and one groyne at Simon colony. The groynes were oriented at about 15° inclination to shore normal from the shore towards the offshore direction and the end of the groyne is curved to take care of the waves coming during Southwest monsoon. It was felt that this proposal would provide shore protection and also help the fisherman to anchor or tow the boats during the months June to September directly from the location of berthing. The effect of the groyne field on adjoining shore was also ascertained by a numerical model that provided the distribution of wave heights due to the presence of groynes. The numerical model also provides the long-term prediction of the shoreline evolution due to the presence of shore-connected groynes and the combined effect of diffraction and refraction due to the presence of groynes and rock outcrops. The groyne field that was proposed for a stretch of 3 km, as shown in Fig. 6.9(b). In certain locations, the natural outcrops are joined by groynes for better tranquillity. The groyne field that was implemented in mid-2004 has been functioning well in trapping the long shore sediment. Due to this measure, beach of width of about 200 m in between the groynes have formed and stabilised which is evident from the Fig. 6.9(c). The groyne field is serving not only as coastal protection measure for the coast of the villages but also act as mini fishing harbours. During the tsunami of 2004, the villages protected

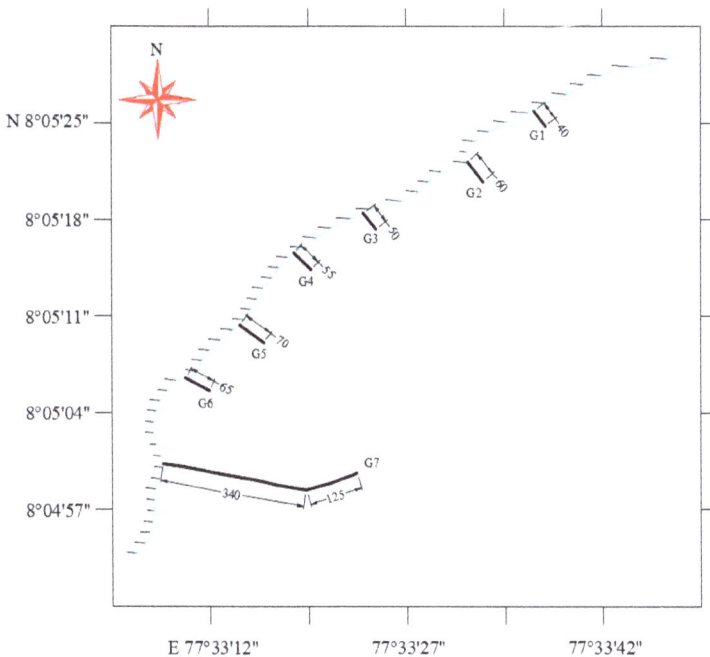

Fig. 6.8(a)   Layout at Kanyakumari.

Fig. 6.8(b)   Shoreline changes at Kanyakumari.

by the beaches formed in between the groynes were least affected, whereas, the adjoining coastal villages did face wrath of the effects of the tsunami. The functioning of this groyne field during the tsunami had paved way for planning of similar ones as it had demonstrated its multifunctional purpose.

6.3.2   *Kerala*

Kerala Coast is a narrow strip of land of length 550 Km, bordering the Arabian sea at the south western part of peninsular India. The coast is characterized by longitudinal barrier strip of low-lying land, sand witched between the Arabian Sea and a continuous chain of lagoons and backwaters

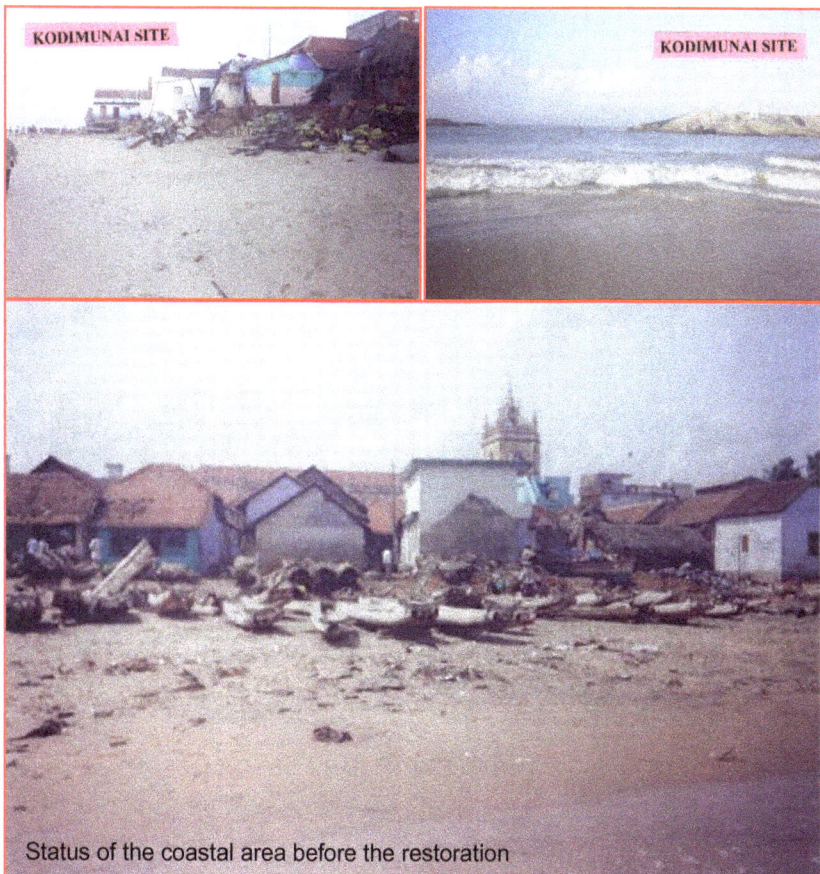

Fig. 6.9(a)   Kodimunai site in the year 2002.

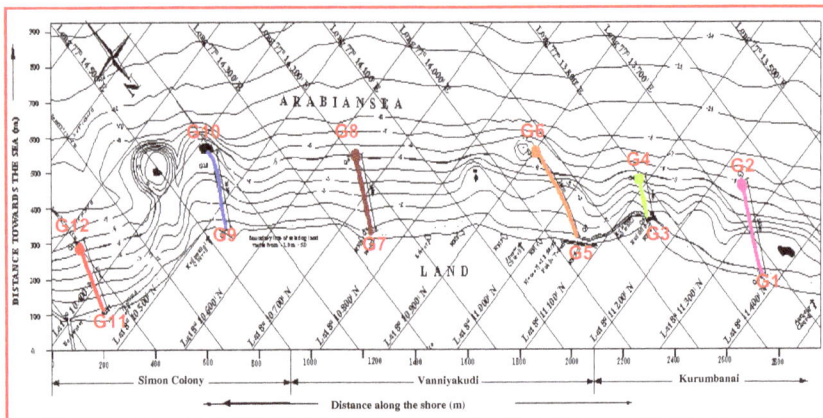

Fig. 6.9(b)   Proposed groyne field.

Fig. 6.9(c)   Ariel view of Kodimunai in Jan 2005.

with connections to the sea at several points. This coastal state experienced considerable damage during the 2004 Indian Ocean Tsunami.

The south and south-central districts of Kerala experienced the diffracted tsunami waves which led to considerable loss of life and property.

N

- - - TSUNAMI AFFECTED AREA

KERALA

ARABIAN SEA

INDIAN OCEAN

Fig. 6.10   Kerala map.

Since the disaster first approached the Tamil Nadu coast, it gave local officials a stipulated time frame of about half hour to evacuate the sensitive stretches of the coast. A few prominent locations of coastal protection along the tsunami affected coastal stretches are discussed below and they are shown in Fig. 6.10.

Design of coastal protection structures in Kerala

The coast of Kerala had been experiencing continuous erosion over several decades. As a protection measure, a single cross section of seawall as shown in Fig. 6.11 was adopted for the affected stretches all along coast. However, this proved to be futile exercise as the sea wall section had failed at several locations and even the traces have disappeared at a few stretches of the coast. The post tsunami signature studies brought to light the extent of damages, which forced to evolve site specific protection

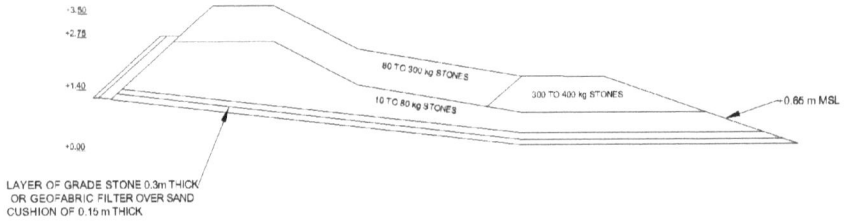

Fig. 6.11   Cross-section of seawall proposed by Kerala irrigation division.

measures. The studies involved comprehensive survey, site specific conceptual designs, physical or numerical modeling, implementation of the scheme followed by monitoring the effectiveness of the protection measures. The typical cross sections of the seawall adopted as per the availability of the beach width to facilitate construction are projected in Table 6.1. This table also includes the location and the year of its implementation.

Panathurakkara Coast (N 8°24′59.8″; E 76°57′49.0″)

The sea wall sections continued to fail ever since its preliminary/primary installation in the year 1974. Post 2005, the Type 1 cross section from Table 6.1, was adopted at the site. The modified cross section withstood cyclonic and other coastal storms without failure and minimum maintenance. The pictures of Panathurakkara coast prior and post implementation of the Type 1 seawall is given in Figs below 6.12(a), 6.12(b), 6.13(a) and 6.13(b).

Poonthura coast (8°26′28.28″N; 76°56′54.42″E)

A piled structure and slab system similar to a skirt breakwater i.e. a semi submerged skirt wall supported by piles was commissioned in 1980, to withstand the fury of waves in front of Poonthura church, the entire structure collapsed due to wave attack in the following monsoon season prior to the structure being completed. During 1981–2005, the original seawall cross section were tried but failed. In 2009–10, a series of groynes based on a site specific study by the authors was undertaken. The groyne field consisted of 8 groynes which were able to trap the sand in between the groynes and the coastal community has become very active in this stretch after 3 decades as can be seen in Fig. 6.14(a), 6.14(b), 6.15(a), 6.15(b), 6.16(a) & 6.16(b). Also on the southern end of Poonthura coast a 20 m long Type 1 seawall was executed and the photos in Fig. 6.17(a) & 6.17(b) show the toe gabions and the impact of sand accumulation after the gabions were placed.

Table 6.1.   Site specific design.

Type	Year of execution	Features	Design cross section
1	2008 Panathura, Trivandrum coast & Tharayilkadavu Allapuzha coast.	Gabion boxes filled with small granite stones, 1–2 Ton stones as armour layer, Base width 26.50 m.	
2	2009 Sraikadu, Kollam, Pathiankar, Allapuzha coast.	Reformation of the existing conventional seawall using geo fabric filter as base layer, core stones and gabion boxes of $1 m \times 1 m \times 1 m$ as armour layer with bas width of 8 m and 12 m	
3	Groyne field Jayanthi colony, Alapad, Kollam district.	4 nos groyne field, 2 nos T-section and 2 nos Bulb section, having bedding layer, core layer, To mound layer, and Armour layer of graded granite stone size varying from location to location	

Fig. 6.12(a)   Damaged seawall at Panathurakkara.

Fig. 6.12(b)   Coast protected using Type-1.

Fig. 6.13(a)   Site specific Type-1 +2 no Groynes Pilot study northern side view towards Poonthura at Panathurakkara.

Fig. 6.13(b)   Site specific Type-1 +2 no Groynes southern side view to Kovalam.

Fig. 6.14(a)   Pile and slab system similar to skirt breakwater.

Fig. 6.14(b)   Eroded coast at Poonthura.

Fig. 6.15(a)   Failed section between damaged piles.

Fig. 6.15(b)   Failed seawall behind pile structure.

Tsunami

Fig. 6.16(a)   Sand accumulation behind piles.

Fig. 6.16(b)   Fisherman community back to normal 2011.

Fig. 6.17(a)   Gabion boxes used as toe wall Type-1 (20 m).

Fig. 6.17(b)   Type-1 in progress Poonthura coast (2012).

Jayanthi colony (9° 6′ 23.0″ N; 76° 28′ 27.8″ E)

During the Indian Ocean Tsunami in Dec 2004, maximum number of casualties occurred at this stretch of the coast, since waves over topped the seawall and washed away houses and 120 lives were reported to be lost. The top level of the seawall was +3.3 m which was not sufficient to prevent overtopping of waves even during the regular monsoon seasons. Hence a series of four groynes were constructed in 2009, and the corresponding cross section of type 3 was adopted. The series of groynes were very effective by way of trapping of sand and for the livelihood activities of the fisherman communities. The proposed layout of the groyne field is shown in Fig. 6.18(a), and the consecutive shoreline changes are shown in Fig. 6.18(b).

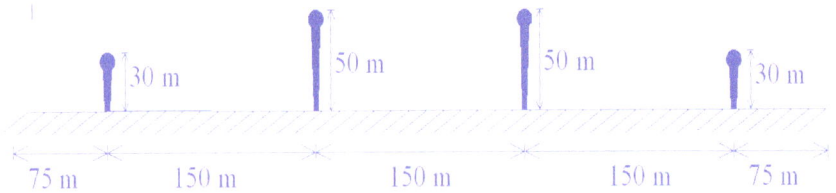

Fig. 6.18(a)   Proposed layout at Jayanthi colony.

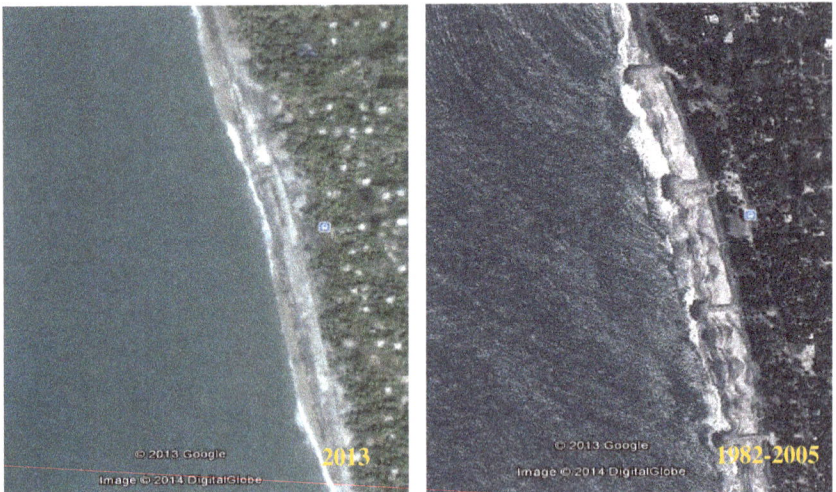

Fig. 6.18(b)   Shoreline changes at Jayanthi colony.

Chennaveli (9° 37′ 52″ N; 76° 17′ 43″ E)

Chennaveli is a coastal hamlet located in the Allapuzha district. The beach area was experiencing extensive erosion over the years and suitable protection measures were mandated to stabilize the coast. A proposal was made to construct 9 groynes along the coastal stretch. The groyne field was composed of transitional groynes, with the longer groynes oriented in the center and smaller groynes towards the ends. The proposed layout is shown in Fig. 6.19(a). The variation in shoreline over the years are shown in Fig. 6.19(b)

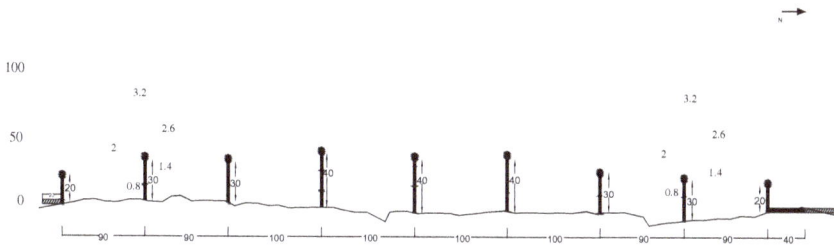

Fig. 6.19(a)   Proposed layout for Chennaveli.

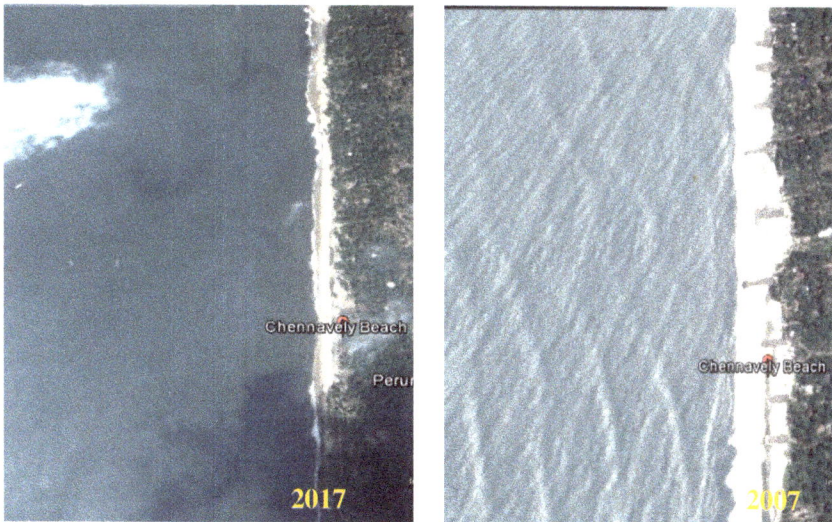

Fig. 6.19(b)   Shoreline variations at Chennaveli.

Perumanathura (8° 43′ 43.79″ N; 76° 42′ 31.64″ E)

The Kadimankulam fishing lake opens to the Arabian Sea at Peru-manathura located in the Thiruvananthapuram district. Owing to the movement of the annual net sediment transportation along the coast, the opening to the lake often gets clogged. Initially, a pair of training walls were constructed with a straight southern training wall and the northern train-ing wall with a bent arm enclosing the approach to the estuary. Sediment bypassing was witnessed from the southern training wall thereby entering inside the mouth. To address the above issue the perpendicular length of the training walls from the shore was increased further and the sand bypass-ing problem was evaded. The bent arm was removed later and a terminal solution was arrived. Figure 6.20 shows the shoreline changes along the coast.

Janarthanapuram, Thiruvanathapura (8° 43′ 43.79″ N; 76° 42′ 31.64″ E)

Varkala is a coastal city and municipality in Thiruvananthapuram district. It is also an important tourist attraction and hence, the shoreline needs to be stabilized without bringing down the natural features. The cliff erosion problem is highly predominant. This coast was protected with submerged geo-tubes, with woven geo-textile founding layer and 2×2 m gabion boxes above the cliff region as shown in Figs. 6.21(a) and 6.21(b). The beach formation in the shore front is shown in Fig. 6.21(c).

Pallana Beach — Allepey district (9° 17′ 55.19″ N; 76° 23′ 18.55″ E)

There are several prototypes of seawall cross section, existing in the world and their workability, efficiency and application are thoroughly site spe-cific. A variety of parameters such as wave climate, soil conditions, tidal range, material availability etc., substantially influence the type of seawall constructed. A conventional seawall is made up of rubble mound, it later evolved to concrete structures, mostly prefabricated units such as kolos, tetrapods, dolos etc., although the design in essence comprises of a core layer, primary and secondary armour layers and toe mound. As opposed to the conventional hard measure of laying rubble stones and concrete units

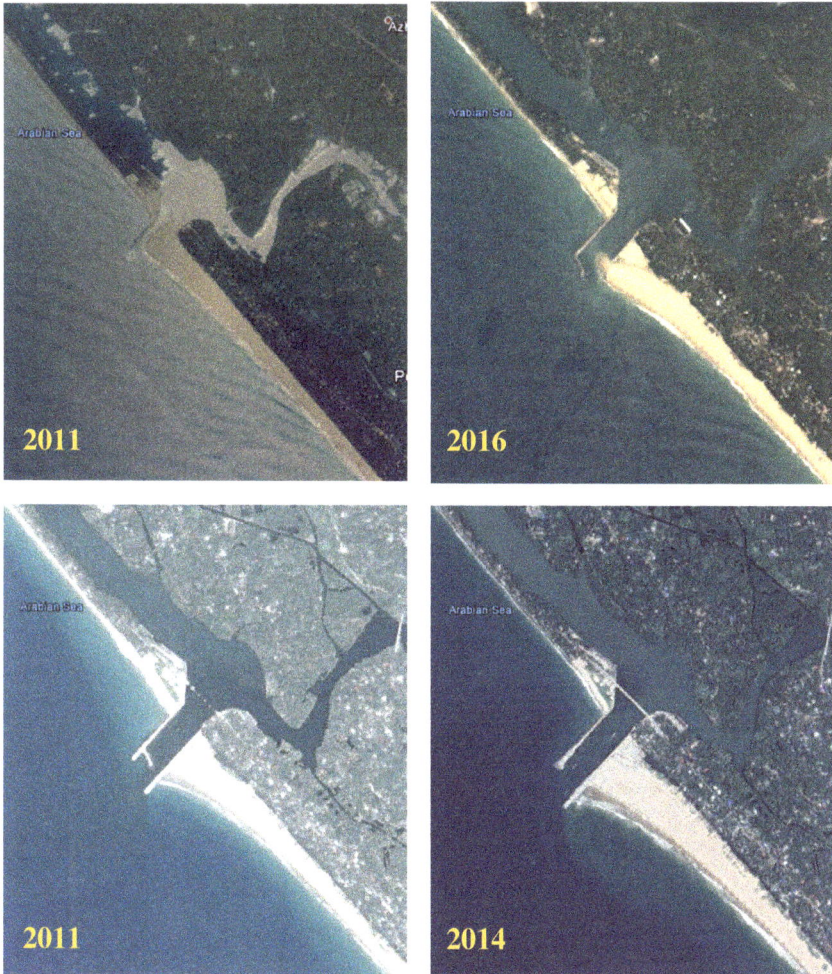

Fig. 6.20   Shoreline changes at Perumanathura.

along the coast; a novel soft measure to erect a seawall using composite geo-synthetic units and locally available beach sand as shown in Fig. 6.22(a) was proposed to be implemented along the Pallana coast of Allepy district in Kerela, India by Flexituff International Limited. A model section of about 100 ft long was installed initially as a test case prototype in the field post

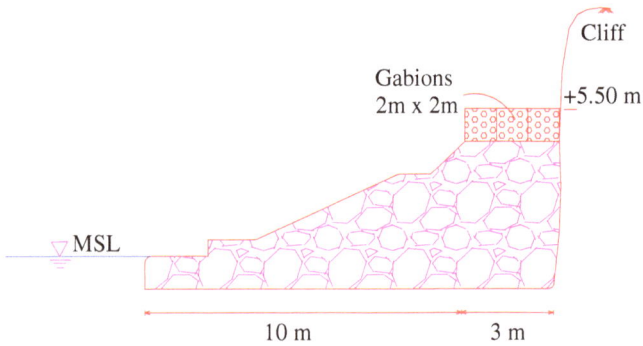

Fig. 6.21(a)   Gabion box type protection near the cliff.

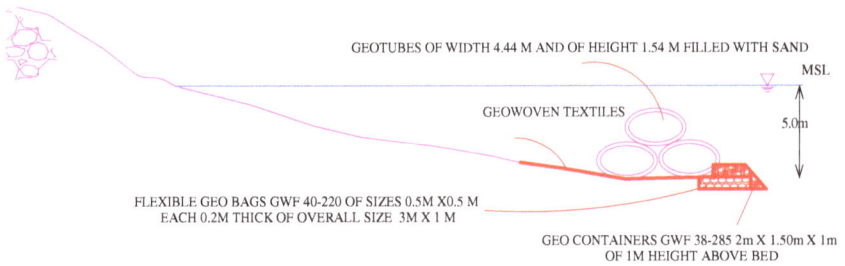

Fig. 6.21(b)   Submerged geo-tube protection.

Fig. 6.21(c)   Shoreline advancements at the Janarthanapuram near Varkala beach.

monsoon, to assess the performance of the seawall in real time. The performance of the geo-synthetic seawall, 3 years after implementation of was assessed and found to be satisfactory even during storms. Figure 6.22(b) shows the images at the site taken during monsoon.

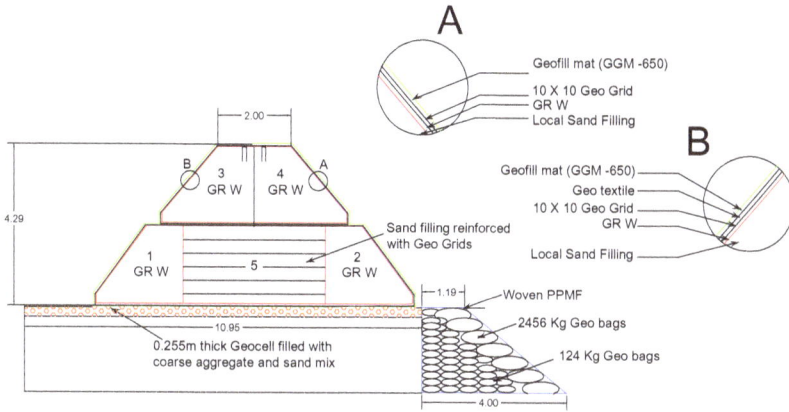

Fig. 6.22(a)   Cross section of the geo-synthetic seawall.

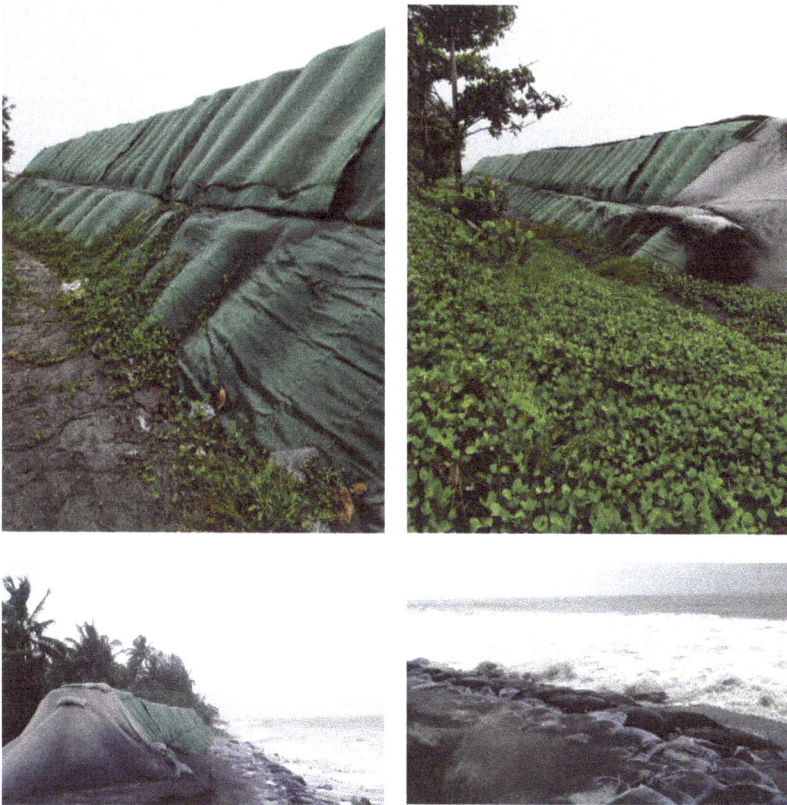

Fig. 6.22(b)   Geo-synthetic seawall at Pallana beach.

6.4   Summary

A general survey of the coast of Tamil Nadu and Kerala after the great Indian Ocean Tsunami was conducted during Feb-March 2005 in order to assess the vulnerable areas being affected by the perennial problem of erosion. The effect of the tsunami was considered in the said exercise. There were two objectives that were considered prior to proposing mitigation plans. The first being identifying the stretches of the coast significantly affected by the Tsunami of 2004. The second being looking at the past records and identifying stretches of the coast affected by perennial problems of erosion, overtopping as well as inundation during extreme events other than tsunami. Prior to proposing the mitigation measures in general, against coastal hazards the studies encompassed preliminary survey, assessment of past records on shoreline changes, assessment of damages during the tsunami, socio-economic aspects of the local community were considered. The proposals consisted of modification of existing structural measures, new conceptual designs backed by detailed physical and numerical modeling and implementation of the schemes. The post-facto information of such implemented protection measures were monitored continuously, the results of which are discussed in detail in this chapter.

References

Sundar, V., and Sundaravadivelu, R. (2005). "Protection measures for Tamil Nadu coast." Report submitted to PWD, Govt. of Tamil Nadu.

Sundar, V., and Murali, K. (2007). "Planning of coastal protection measures along Kerala coast." Report submitted to Govt. of Kerala.

PART 3
Physical Modelling

Chapter 7

Tsunami Detection

7.1 Introduction

An approaching tsunami is detected using various techniques, which include recognition of a high impact earthquake originating adjacent to a critical fault line in the seabed, drastic/notable recession in sea level and other technology-based warning systems based on environmental sensors and data processing. This chapter reviews and evaluates the detection methodologies with special attention to techniques using a surface-floating buoy studied under controlled laboratory conditions.

7.2 Tsunami Warning

The initial warnings after an earthquake are based on the analyses of data from a global seismic detection network, in conjunction with the historical record of tsunami events, if any, at the different seismic zones (Weinstein, 2008). Although adequate for most medium-sized earthquakes, in the case of very large earthquakes or tsunami earthquakes, the initial seismological assessment can underestimate the earthquake magnitude and lead to errors in assessing the tsunami potential. Far from the tsunami source, data from sea-level networks provide the only rapid means to verify the existence of a tsunami and to calibrate numerical models that forecast the subsequent evolution of the tsunami. The current global seismic network is adequate and sufficiently reliable for the purposes of detecting likely tsunami-producing earthquakes by analysing its critical data stream. The tsunami detection and forecasting process requires real-time observations of tsunamis from *both* coastal sea-level gauges and open-ocean sensors (such as provided by the Deep-ocean Assessment and Reporting of Tsunamis (DART) network) as discussed in Section 1.9.

As described in the subsequent sections, the rapid detection of a tsunami striking within minutes to an hour, either for the purpose of providing an initial warning or for confirming any natural warnings that near-field

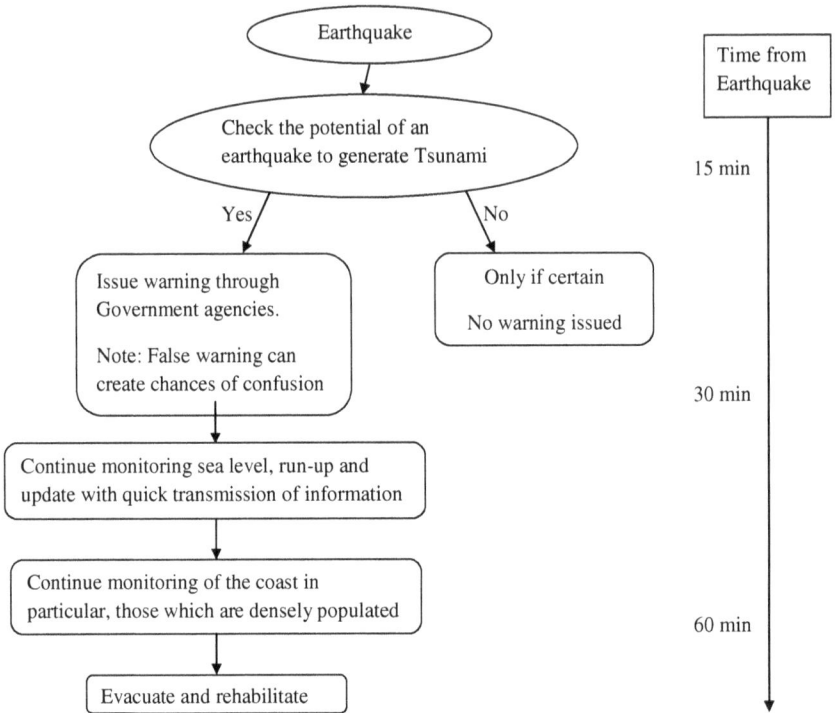

Fig. 7.1   Tsunami warning flowchart.

communities have already received, will likely require consideration of alternative detection technologies, such as sensors deployed along undersea cabled observatories and coastal radars that can detect a tsunami's surface currents tens of kilometres from the shore. A schematic representation of the Tsunami Warning System is discussed in Fig. 7.1.

7.3   Detection of Tsunamis with Sea-Level Sensors

Since the seismic signal is the first observation available to the Tsunami Warning Centers (TWCs), seismic detection provides the basis for the initial evaluation of the potential for a tsunami. The decision about the content of the first message from the TWCs is based solely on seismic parameters and the historical record, if any, of tsunamis emanating from the neighbourhood of an earthquake. However, as previously noted, this indirect seismic method is limited in its accuracy on the strength of the tsunami, usually underestimating its potential of large earthquakes as well as tsunami earthquakes. In acknowledgement of this bias, and because forecasters must err on the side of caution when human lives may be at stake, the TWCs adopt a conservative criteria to trigger advisories, watches or warnings based on this initial seismic assessment (e.g., Weinstein, 2008). However, these conservative assessments might cause unwarranted evacuations, which lead to threat to life and monetary loss. A TWC must, therefore, not only provide timely warning of a destructive tsunami but must also avoid causing unnecessary evacuations with their attendant negative impacts.

The detection and forecasting process requires real-time observation of the tsunamis from both coastal sea-level gauges and open-ocean sensors (such as provided by the DART stations). The combination of the open-ocean and coastal sea-level stations, which provide direct observations of tsunami waves, are important for adjusting and cancelling warnings as well as for post-tsunami validation of models of the tsunami propagation and inundation (U.S. Indian Ocean Tsunami Warning System Program, 2007). These sea-level networks can also detect tsunamis from sources that fail to generate seismic waves or are generated by an earthquake on land that generates a sub-aerial and/or a seafloor landslide. Progress to expand the ocean observing network and advances in oceanographic observing technologies allow the TWCs to incorporate the direct oceanographic detection of tsunamis into their decision processes. The future of tsunami detection is based on satellite observations.

7.4 Indian Tsunami Early Warning System (ITEWS)

ITEWS is governed by INCOIS, an autonomous body under the Ministry of Earth Sciences, Government of India. It actively monitors an earthquake event whenever it is recorded with magnitudes greater than or equal to 6.5 within the Indian Ocean and magnitudes greater than 8.0 outside the Indian Ocean. Parameters such as the earthquake's epicentre, magnitude, origin time and depth are assessed immediately.

The initial bulletin provides information regarding the preliminary earthquake information and its tsunami potential. From the pre-run model database, initial predictions for the matching scenario are selected based on the earthquake parameters. If pre-run model scenario indicates Estimated Wave Amplitude (EWA) <0.2 m then Type-II is issued with NO THREAT information. However, the monitoring of sea-level observations continues if the EWA is greater than 0.2 m, and then Type-II is issued with

Table 7.1.   Threat level status criteria.

Pre-run Model Scenario Results			
ETA < 60 min		ETA > 60 min	
EWA (m)	Threat Status	EWA (m)	Threat Status
> 2	Warning	> 2	Alert
0.5–2	Alert	0.5–2	Watch
0.2–0.5	Watch	0.2–0.5	Watch

Estimated Time of Wave Arrival (ETA), Estimated Maximum Wave Amplitude (EMWA) and Threat Category (WARNING/ALERT/WATCH) for each of the coastal forecast zones.

The total available warning time (i.e., predicted time for the tsunami to reach the coast) available dictates the type of threat (WARNING/ALERT/WATCH) to be issued for a particular region. The coastal areas that lie within less than 60 min travel time should be warned based on earthquake information only, owing to the lack of time to analyse data from bottom pressure recorder (BPR) and tide gauges. The regions beyond 60 min of a potential tsunami attack can be classified under alert/watch and can be later updated to warning after analysing the water level data. Table 7.1 shows the various criteria for issuing threat status.

7.4.1   Importance of floating data buoys

The detection system of a likely tsunami essentially comprises a fixed bottom-mounted pressure transducer and a surface-floating data buoy. The BPR, at the seabed level does not undergo any significant dynamic changes in its position, whereas the motion characteristics of the surface-floating buoy is required to be studied/analysed on the variations it undergoes owing to the action of ocean waves. These surface-floating structures should be capable of staying afloat and transmitting wave conditions. Thus, detailed experimental studies were carried out in order to assess the motion characteristics of the buoy.

7.5   Experimental Studies

7.5.1   General

In order to collect the Metocean parameters and to monitor the marine environment over Indian Seas, the Ministry of Earth Sciences, Government of India has established a National Data Buoy Program (NDBP).

Under this programme, 25 discus data buoys are deployed in Bay of Bengal and Arabian Sea to measure and supply data to improve the weather and ocean-state prediction in Indian waters. All the data buoys deployed under NDBP are 2.2 m diameter discus-shaped buoy hull, which is a slightly modified version of 3 m discus buoys that are widely adopted worldwide. These data buoys are deployed in shallow and deep-water depths, the locations of which are shown in Fig. 7.2. These data buoys collect various oceanographic and meteorological parameters that are essential for the continuous monitoring of the coastal and marine environment along the Indian coast. In the event of natural disaster situations, such as cyclones or tsunamis, the meteorological data obtained from these buoys are vital for effective mitigation measures. In addition, the oceanographic data collected are useful for the understanding of ocean circulation. Apart from these, various other parameters directly or indirectly are used to identify the potential fishing zones, to develop predicting models, to design marine structures for better navigation aid. The 2.2 m diameter discus-shaped buoy of NDBP, shown in Fig. 7.3, is selected for the present experimental study and scale modelled

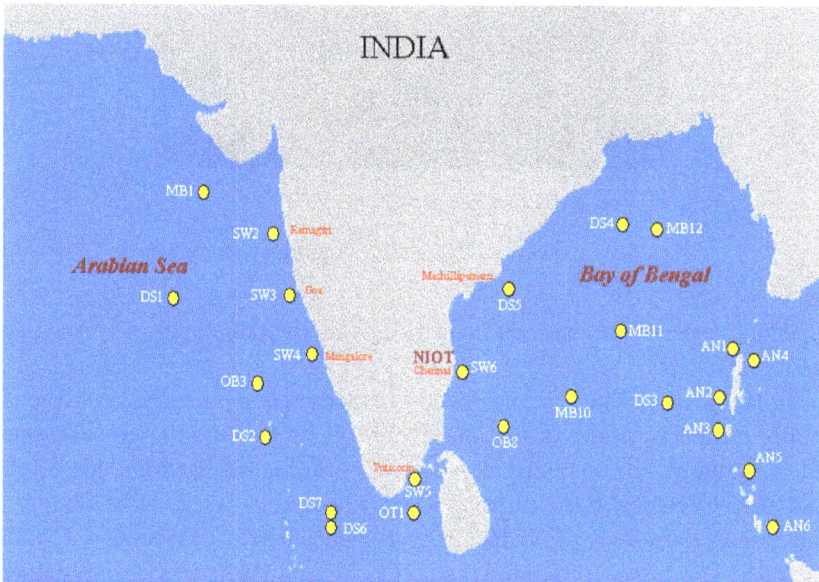

Source: www.niot.res.in.

Fig. 7.2   Buoy locations around the Indian waters.

Source: www.niot.res.in.

Fig. 7.3   View of NDBP data buoy under operation.

to examine its effectiveness in identifying the different types of waves that are experienced.

In the present chapter, the motion response characteristics of a discus-shaped data buoy under the action of solitary and Cnoidal waves are investigated through well-controlled experimental programme (Balaji *et al.*, 2006). The characteristics of the said waves are claimed to be closer to that of a tsunami. A data buoy of NDBP (Government of India) was selected, scale modelled and tested. The model selected for the study is a discus-shaped hull data buoy. The diameter of the discus buoy is 2.2 m with a height of 1.07 m and weighs of about 950 kg. Adopting a model scale (λ) of 8, the dimensions of the buoy is scaled down, the details of which are given

Table 7.2.   Details of the model.

Description	Prototype	Dividing factor	Model
Diameter	2.2 m	λ	0.275 m
Height	1.07 m	λ	0.134 m
Weight	950 kg	λ^3	1.86 kg

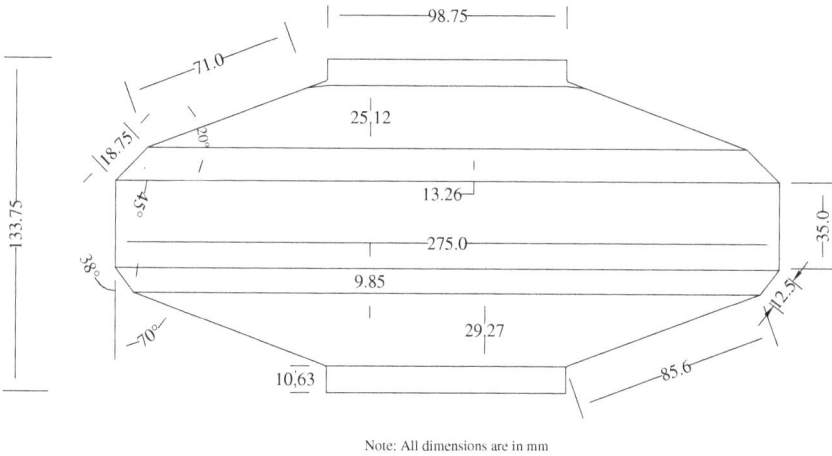

Note: All dimensions are in mm

Fig. 7.4   Cross-sectional details of the buoy model.

in Table 7.2. The cross-sectional details of the buoy model are given in Fig. 7.4.

The tests were carried out in a wave flume 30 m long, 2 m wide and 1.7 m deep. The water depths were varied from 0.2 to 0.25 m. Infra-red cameras emit infra-red light, which is reflected off low mass, retro-reflective markers and back to the camera to generate a 2-dimensional image of the marker positions. By triangulation with the signals from more than one camera, a three-dimensional recording of the positions of the marker can be generated. In the present study, two MCUs were used to calculate the positions of the buoy model. The camera uses the reflected data to calculate the position of the targets with a high spatial resolution. ProReflex MCU uses charge-coupled device (CCD) image sensor to capture the reflection from the markers with a resolution of 658 × 500 pixels with a maximum measuring frequency range of 120 Hz. The details of these experiments are provided in detail by Balaji (2007).

7.5.2   Cnoidal waves

7.5.2.1   General

Cnoidal waves of period ranging from 2.5 s to 3.0 s with height varying from 0.24 m to 0.31 m were simulated for the present study. The range of Ursell parameter $(U_r = HL^2/d^3)$ 25.8–38.4 and a range of d/L of 0.105–0.12 were covered. Hence, the conditions for the validity of Cnoidal wave theory (a) $d/L > 0.125$ and, (b) $U_r > 26$ were satisfied.

The measured Cnoidal wave elevation, in general, is found to be in good agreement with that of the theory as shown in Fig. 7.5 for two different U_r parameters. The Cnoidal wave elevation and the corresponding buoy responses were simultaneously measured and typical plots of which along with their respective frequency spectrum is shown in Fig. 7.6. It is observed from these frequency spectra of the buoy responses that they contain more than one spectral peak. Interestingly, the spectral peaks appear at the integer multiples of the fundamental frequency (f_0). This is due to the fact that the Cnoidal wave elevations are composed of fundamental period and its integer multiples (higher harmonics). Although the magnitudes of these higher harmonics are less in the wave spectrum, the effect is significantly felt in the response spectra. This motivates the analysis of these time histories in the phase projections, spectral and harmonic analyses.

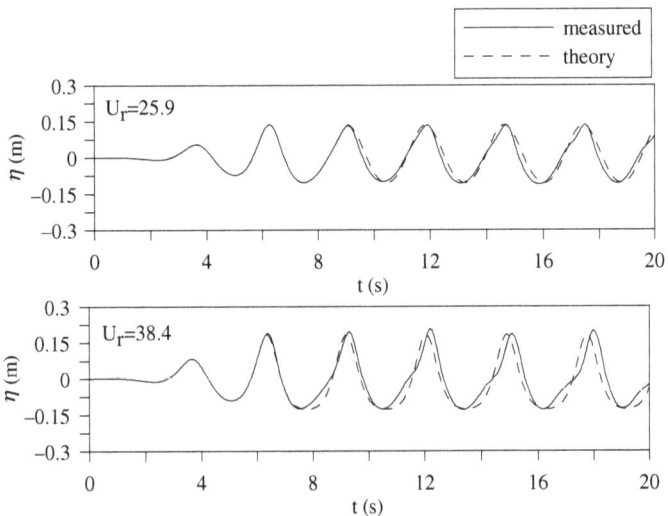

Fig. 7.5   Typical comparison of measured and theoretical Cnoidal wave elevations.

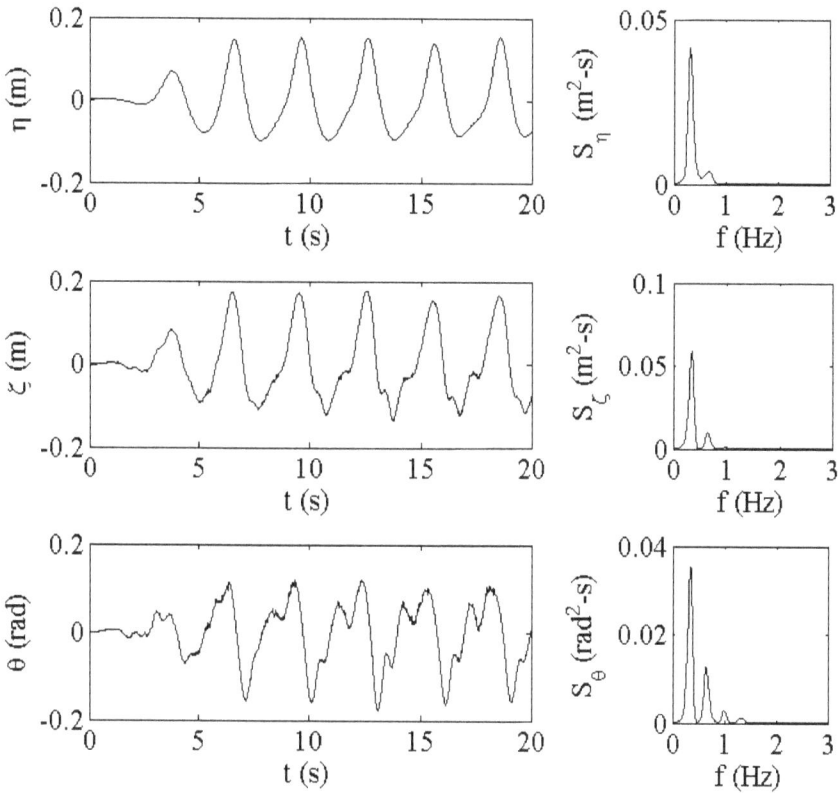

Fig. 7.6 Typical measured time histories and their frequency spectra ($U_r = 30.9$).

7.5.2.2 *Effect of wave parameter on the responses*

The measured Cnoidal wave elevation and the respective responses are plotted against their respective time derivatives. Such phase portraits for a Cnoidal wave height H of 0.31 m are shown in Fig. 7.7(a) and 7.7(b). The shape of the phase portrait for the wave elevation is non-symmetrical indicating that the crests are sharper than the troughs, an aspect that the physical profile of the Cnoidal wave elevations exhibits. In general, the size of the phase portraits for the response increases with an increase in U_r due to the increase in the wave period and subsequent incident energy. The multiple period attractor seen in the phase portraits of responses are due to the existence of more than one frequency component in the time histories. An increase in the U_r leads to the complex shape of the response

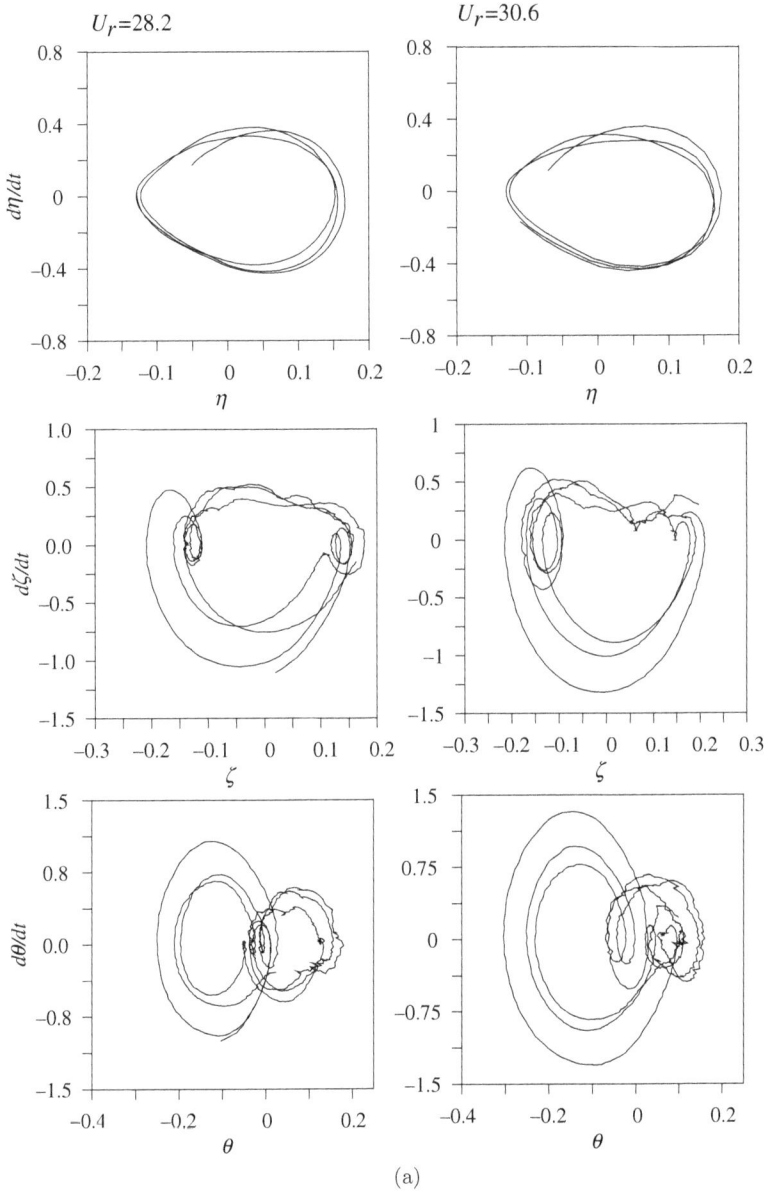

Fig. 7.7   (a) Effect of Cnoidal wave periods on the variations of phase portraits.
(b) Effect of Cnoidal wave periods on the variations of phase portraits.

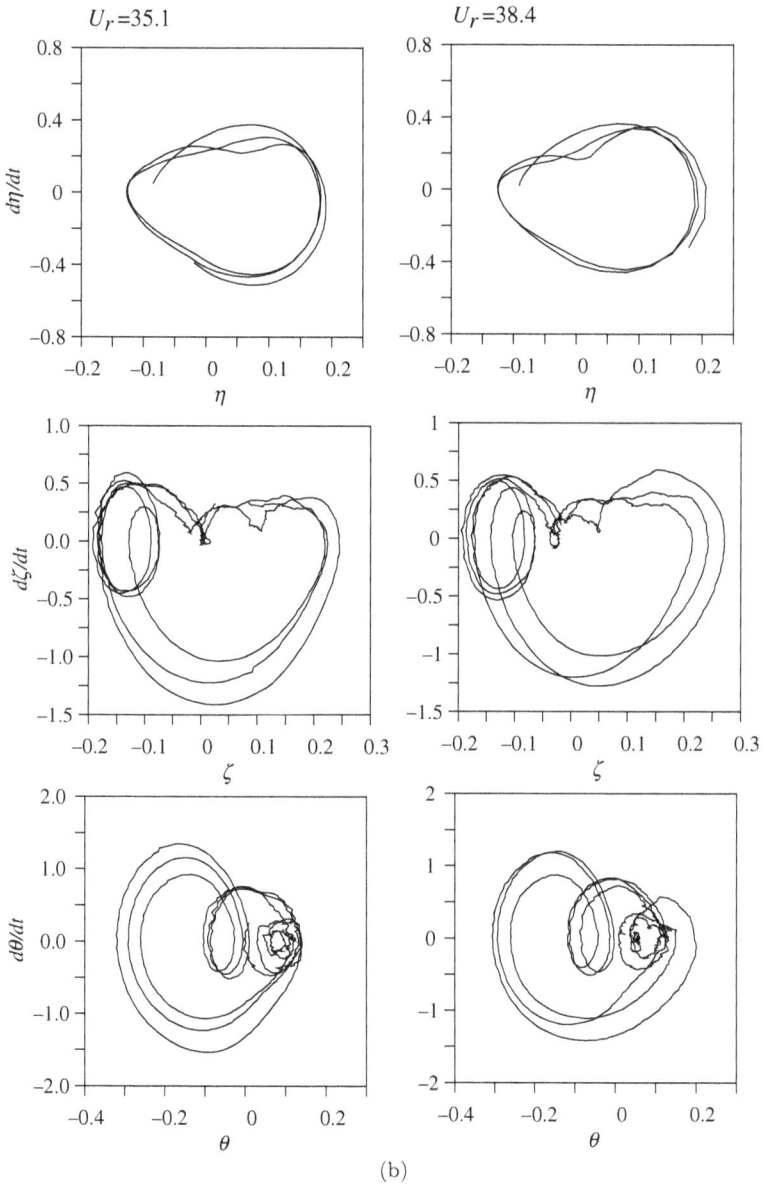

(b)

Fig. 7.7 (*Continued*)

phase portraits. The effect of the Cnoidal wave height on the variations
of the phase portraits for two different wave periods, T 2.7 s and 2.8 s, are
shown in Fig. 7.8(a) and 7.8(b), respectively. It is clear from the results
that variation in the wave height does not influence the complexity of the
phase portraits, but the magnitude is found to be increased.

The wave elevation and the buoy model responses were subjected to
spectral analysis. The frequency spectra of wave elevation and responses are
plotted against non-dimensional frequency for two wave heights of 0.31 m
and 0.24 m in Fig. 7.9(a) and 7.9(b), respectively. It is observed from the
results that the spectral peaks appear at the integer multiples of the fundamental frequency f_0. The spectral density for all the higher harmonics
is found to increase with an increase in U_r, due to an increase in the nonlinearity. The peaks appear to shift from the exact multiples of f_0 with
an increase in U_r. The spectral peaks shift towards the left of the integer multiple with an increase in U_r for higher wave height, whereas the
trend is reverse for the lesser wave height. As the wave height increases,
the nonlinear wave propagation leads to transfer of energy to lower frequency components in the form of nonlinear wave–wave interaction. This
leads to a shift in the peak frequency towards lower frequency zone. The
corresponding integer multiples of the fundamental frequency components
is found to also shift towards lower frequency region. Interestingly, all the
spectral peaks of higher harmonics appear closely at the multiple integer of
f_0 for an increase in the wave heights for a constant wave period. The wave
and response spectra for two different wave periods, T of 2.7 s and 2.8 s are
presented in Fig. 7.10(a) and 7.10(b), respectively. It is understood from
the results that the higher harmonics are significantly influenced by the
change in the Cnoidal wave period.

The normalized average spectral density of heave and pitch responses at
the higher harmonics are plotted with respect to U_r for two constant wave
heights in Fig. 7.11. The spectral density of responses at higher harmonics
increases with an increase in the U_r for both the wave heights. Among the
responses, the pitch spectral density drastically increases with respect to an
increase in U_r and a maximum of about 58% is observed at second harmonics for a U_r of 38.4 tested. The corresponding value for the heave response is
20%. With an increase in the higher harmonics, the non-dimensional spectral density decreases for all the U_r and reduces to the smallest value of
3% for the pitch responses at $f = 5f_0$. The effect of change in the Cnoidal
wave heights on the non-dimensional spectral density at higher harmonics

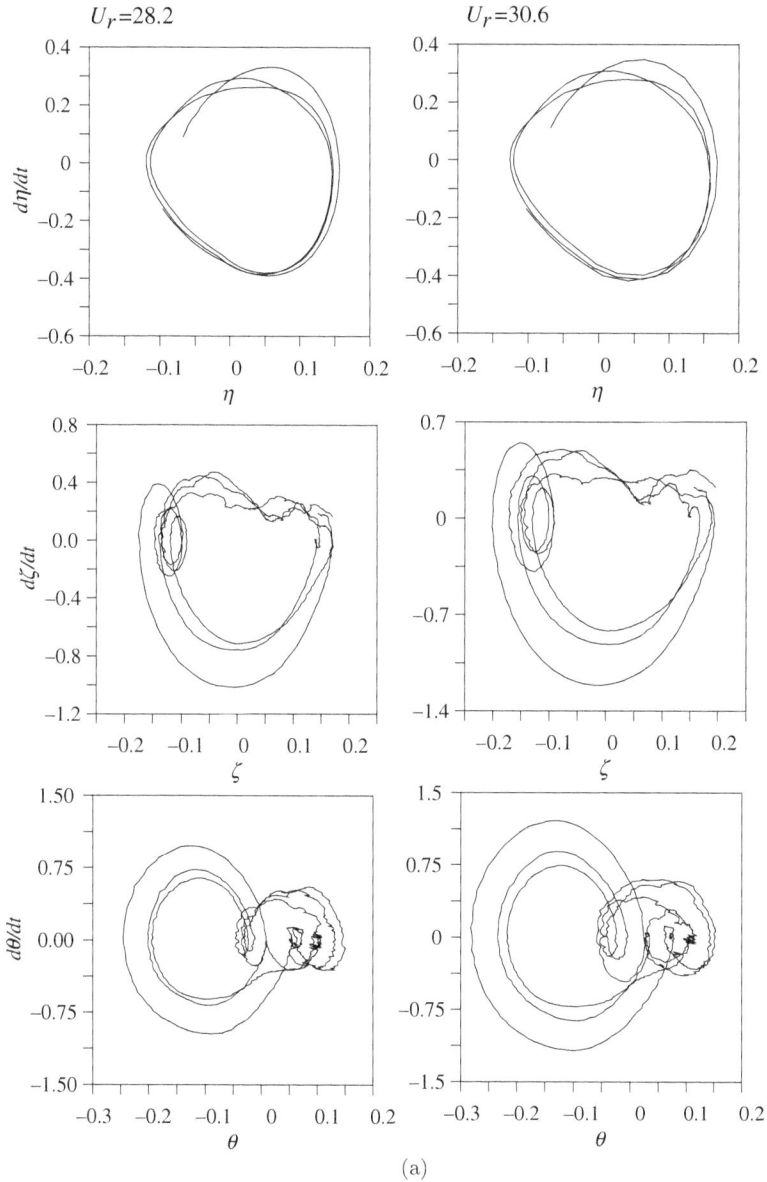

Fig. 7.8   (a) Effect of Cnoidal wave heights on the variations of phase portraits.
(b) Effect of Cnoidal wave heights on the variations of phase portraits.

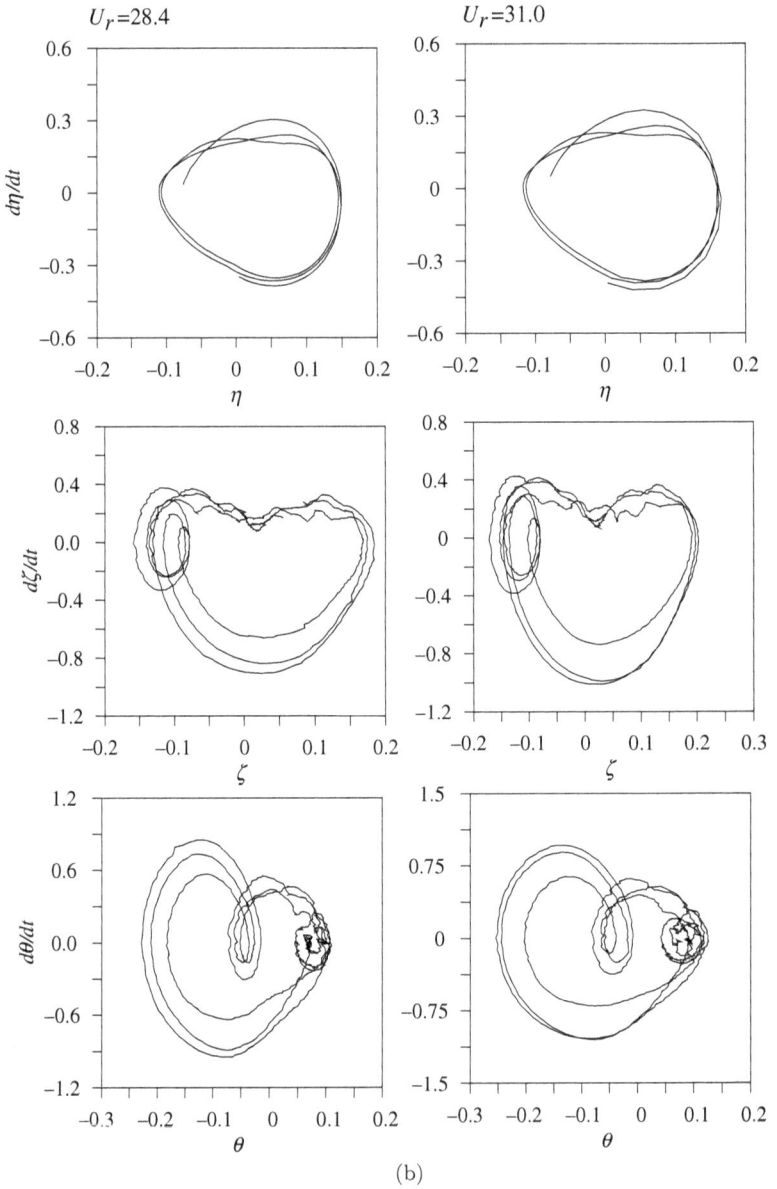

Fig. 7.8 (*Continued*)

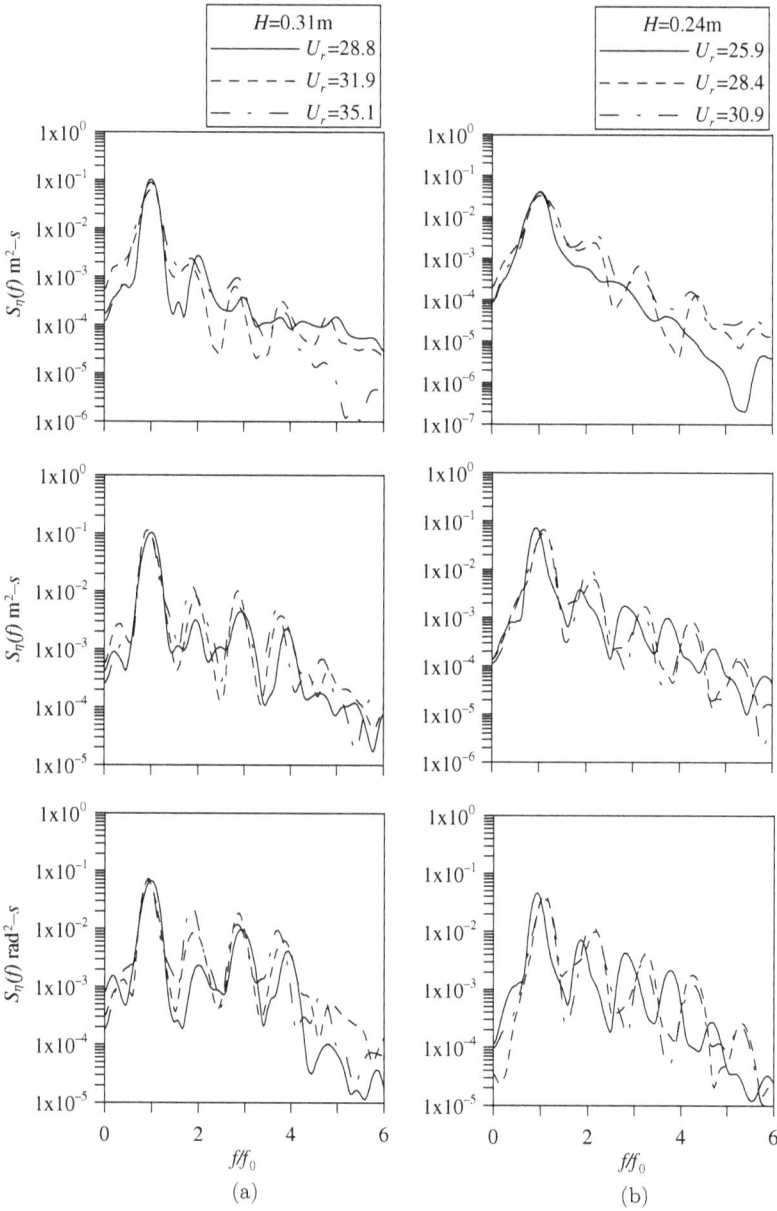

Fig. 7.9   Effect of Cnoidal wave period on the spectral energy variations.

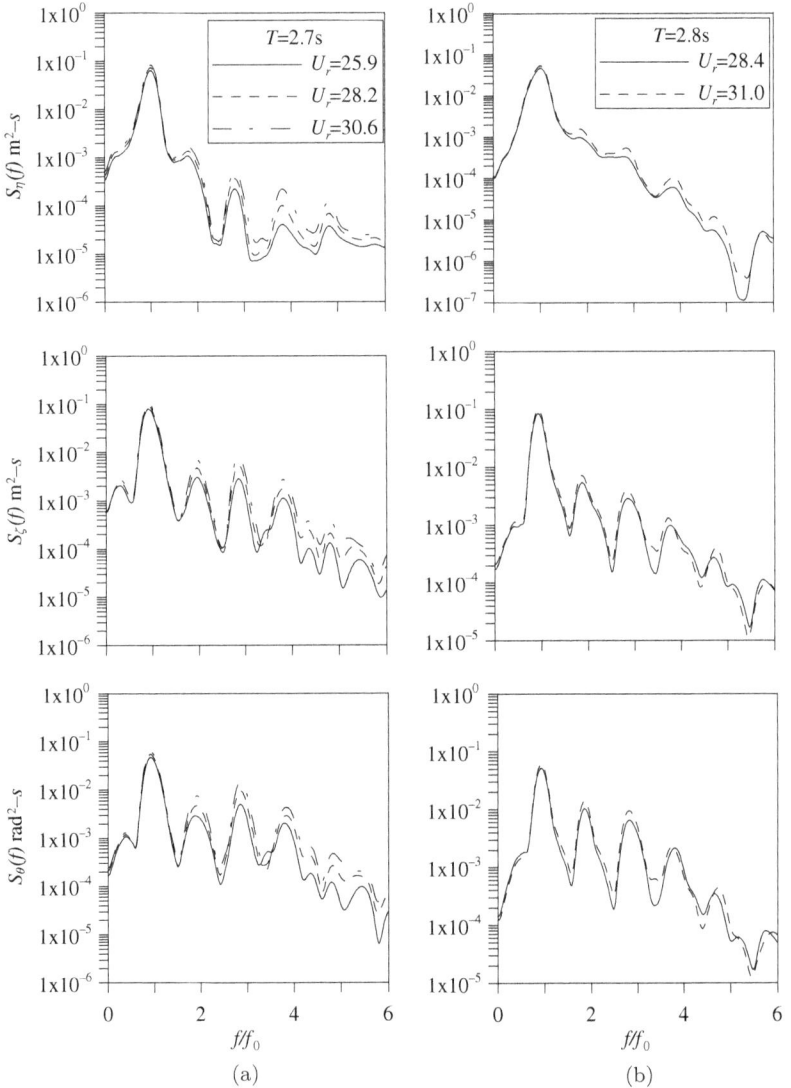

Fig. 7.10   Effect of Cnoidal wave height on the spectral energy variations.

is shown in Fig. 7.12 for two different wave periods. The variation of energy levels with respect to U_r is similar to that discussed earlier. The heave and pitch response spectral density at higher harmonics decreases with decrease in T, as the wave periods decreases and so do the nonlinearity.

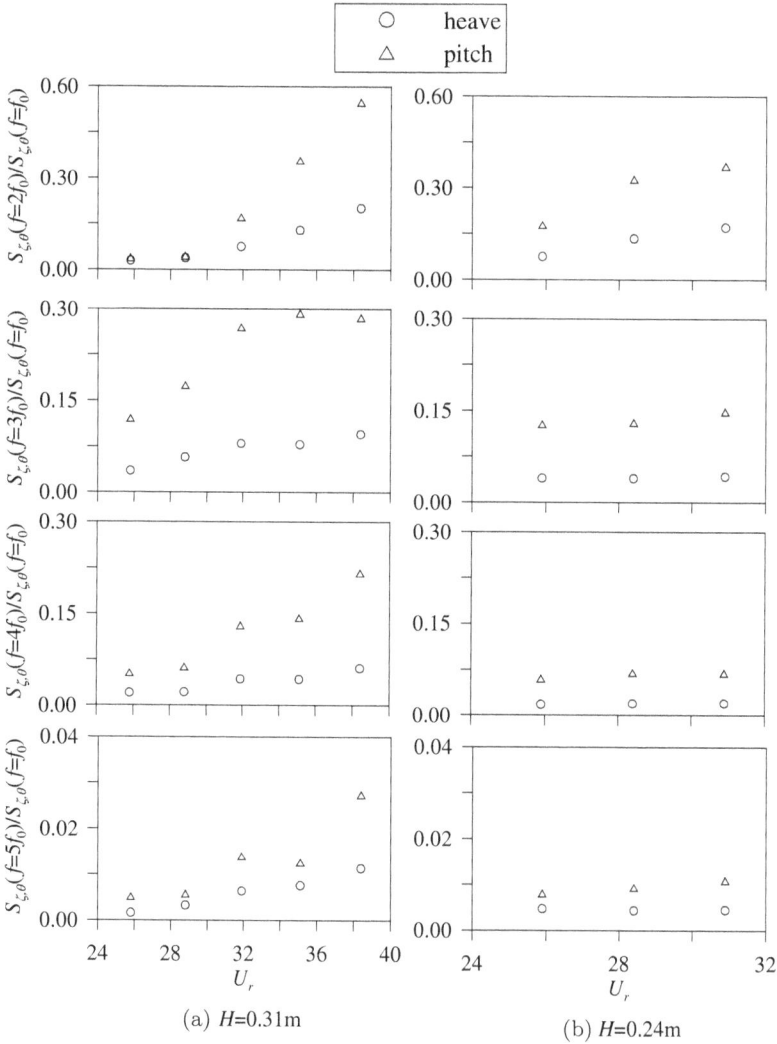

Fig. 7.11   Effect of Cnoidal wave period on the spectral energy variations at harmonics.

7.5.2.3   *Prediction using transfer function, TF*

By adopting the harmonic transfer function (HTF) established in the preliminary studies, the Cnoidal wave spectral densities are predicted from the buoy model response spectra. The comparison of measured and predicted wave elevation spectra for a constant H of 0.31 m is shown in Fig. 7.13.

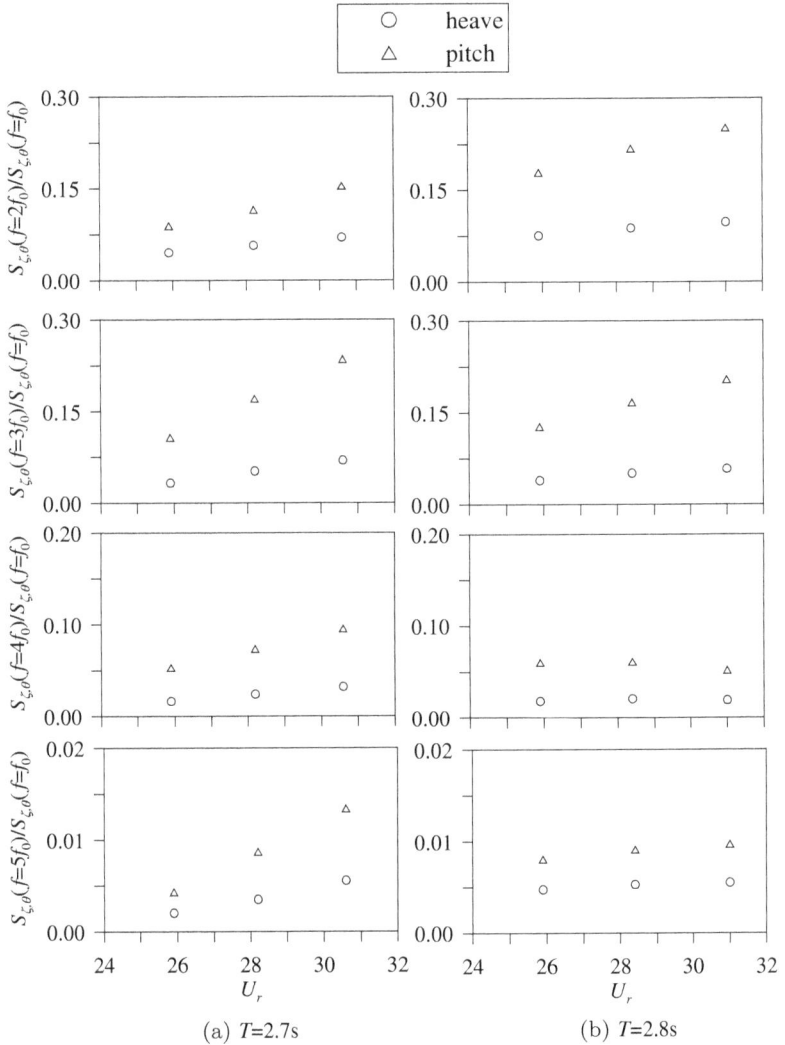

Fig. 7.12   Effect of Cnoidal wave height on the spectral energy variations at harmonics.

Similarly, the comparison for a constant T of 2.7 s is shown in Fig. 7.14. It is clearly observed from these results that the predictions are always overestimated, due to the fact that the heave energy of the buoy under the Cnoidal wave conditions is more than that of under the normal monochromatic conditions. The percentage overestimation of the wave spectral density near

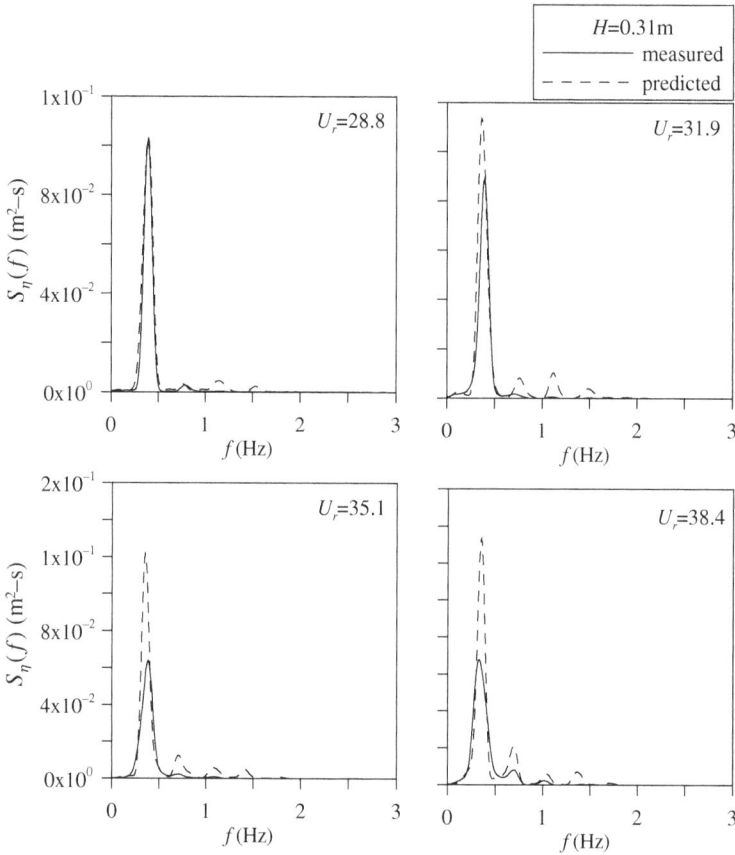

Fig. 7.13   Prediction using HTF — Effect of Cnoidal wave period.

the peak frequency is calculated for the entire tested range of Cnoidal wave conditions and plotted against U_r in Fig. 7.15, for typical constant H and T. Although there is an overestimation, an increase in the Cnoidal wave height did not affect the estimation of wave spectral density, further. On the other hand, an increase in the Cnoidal wave period for constant wave heights significantly influences the predictions of wave spectral densities. For the highest U_r tested, a maximum percentage overestimation of about 120% is observed within the tested range of Cnoidal wave conditions. The estimation of nonlinear wave spectral densities using the linear TF would certainly lead a higher estimation of energies which obviously affect the design considerations of any marine structures in the shallow water depths.

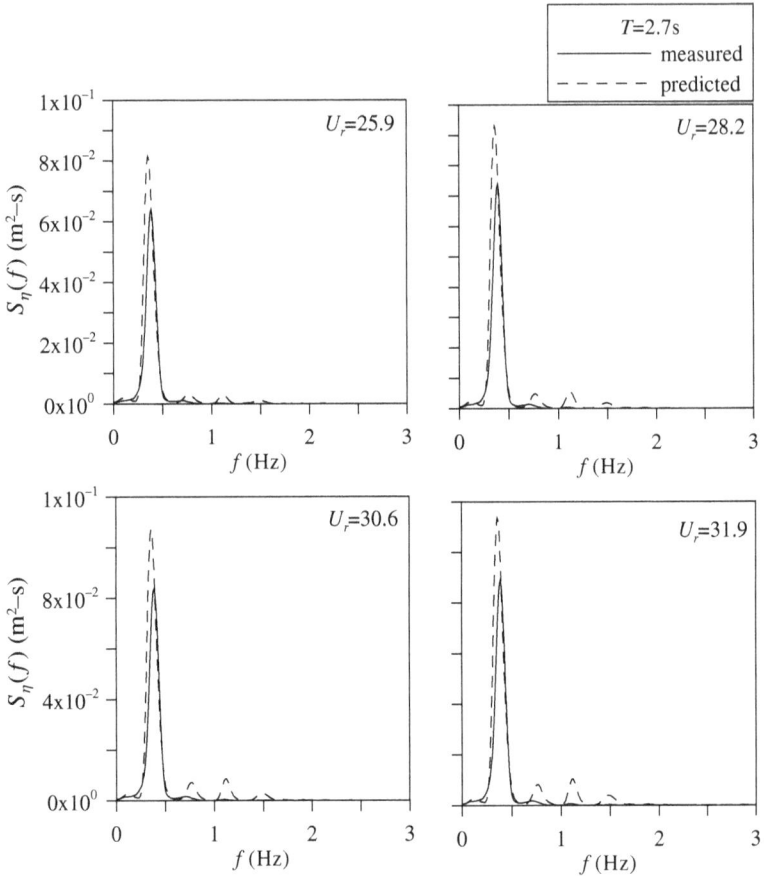

Fig. 7.14   Prediction using HTF — Effect of Cnoidal wave height.

7.5.3   *Solitary waves*

7.5.3.1   *General*

For the present study, solitary waves with heights ranging from 0.04 m to 0.1 m were adopted, such that (H/d) varied from 0.05 to 0.125. It is to be noted that for a water depth of 0.8 m, due to the limitations of the wavemaker motions, the maximum possible solitary wave height that could be generated in the wave flume is 0.1 m.

The comparisons of measured and theoretical solitary wave elevations, in general, are found to be in good agreement, as can be seen in Fig. 7.16 for

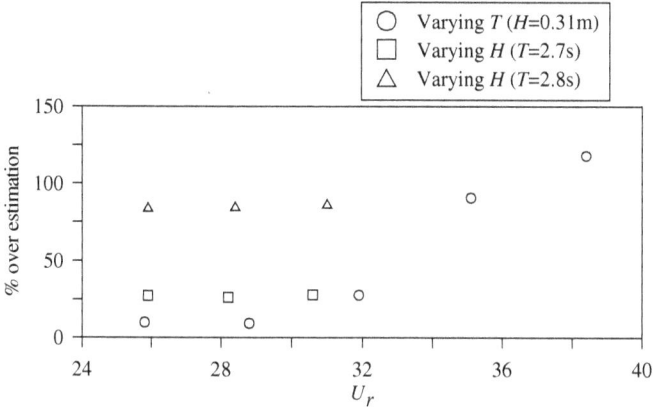

Fig. 7.15   Percentage overestimation using HTF.

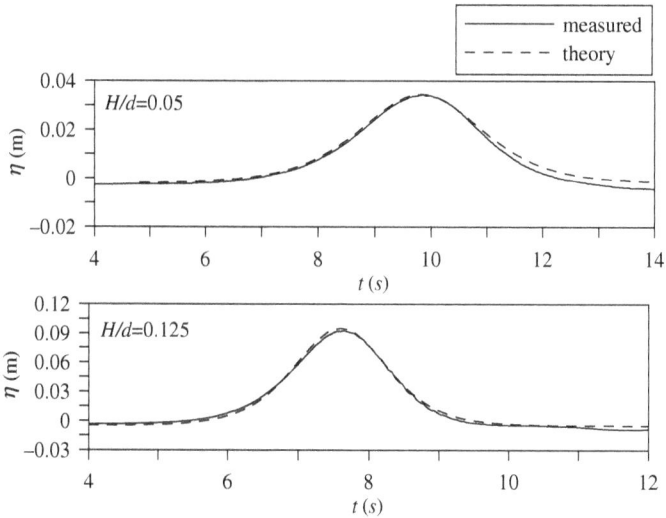

Fig. 7.16   Typical comparison of measured and theoretical solitary wave elevations.

two different H/d values. The solitary wave elevation and responses were simultaneously measured and typical plots of which are shown in Fig. 7.17. The heave response of the buoy model closely follows the solitary wave profile, whereas the pitch oscillates about its natural period.

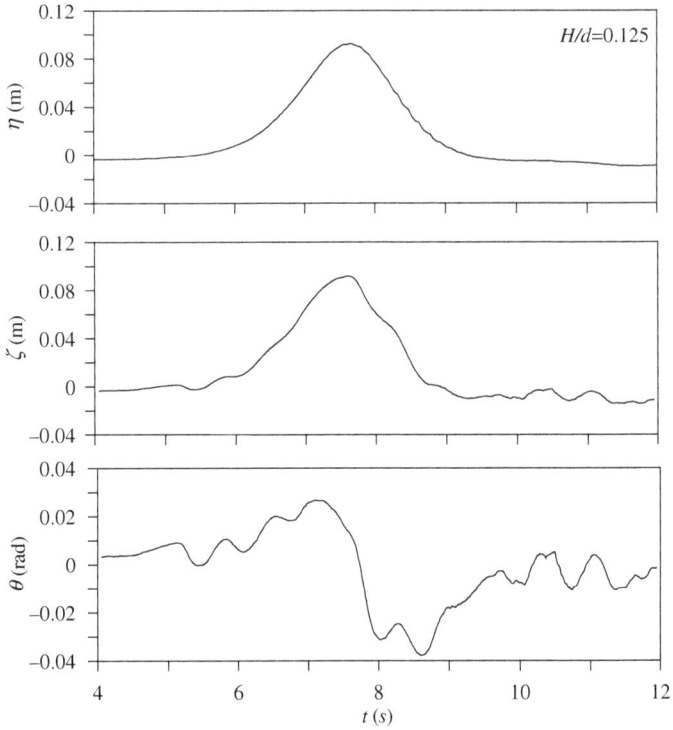

Fig. 7.17   Typical measured time histories.

7.5.3.2   *Effect of wave parameter on the responses*

The phase portraits of solitary wave elevation and the response for two different H/d values are shown in Fig. 7.18. The phase portraits of wave elevations fall only on the positive side of the x-axis, exhibiting the presence of only a single crest in the wave elevation time history. Although a similar trend in the variation of the phase portraits is observed for the heave, more than one period attractor is seen, which may be due to the natural period of the buoy model in the heave mode. The presence of both positive and negative values in the phase portraits of pitch response indicates that the buoy model slides towards the front and rear side of solitary wave profile. With an increase in H/d, the magnitude of the phase portraits of pitch increases and the period attractor is more clearly seen.

The measured time histories of wave elevation and response were subjected to spectral analysis and the spectral densities are plotted for different

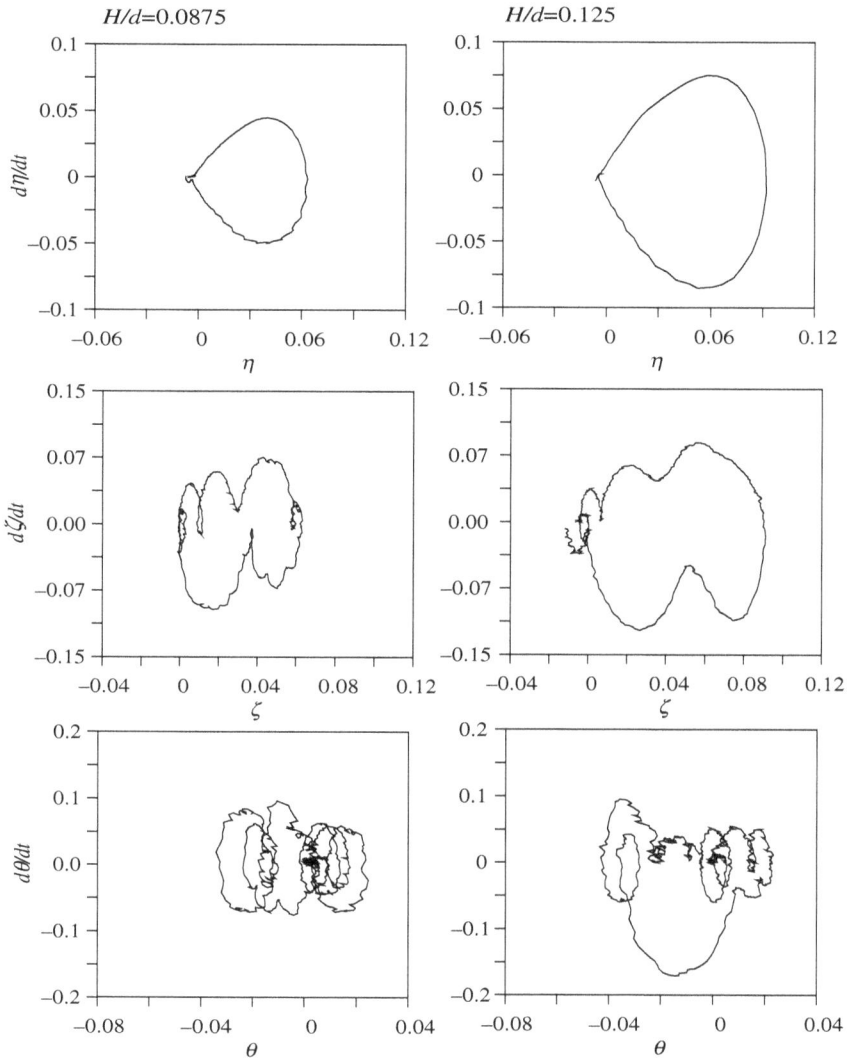

Fig. 7.18 Phase portraits of solitary wave and responses.

H/d values as shown in Fig. 7.19. The spectral densities of wave elevation and the respective response are found to be higher for larger H/d, due to an increase in the incident wave energy. The energies of the response fall in the low frequency region, which correspond to the solitary wave frequencies. The spectral peak of heave and pitch is seen to occur around the peak

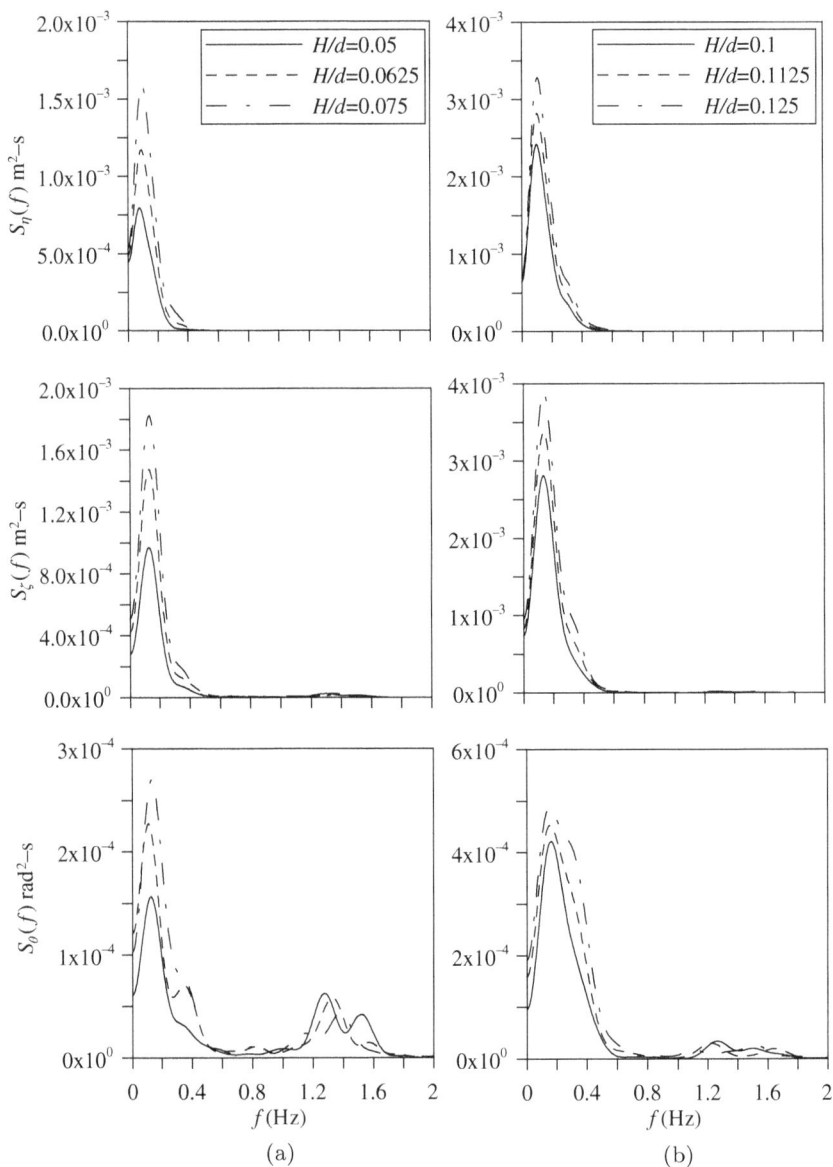

Fig. 7.19   Frequency spectra of solitary wave elevations and responses.

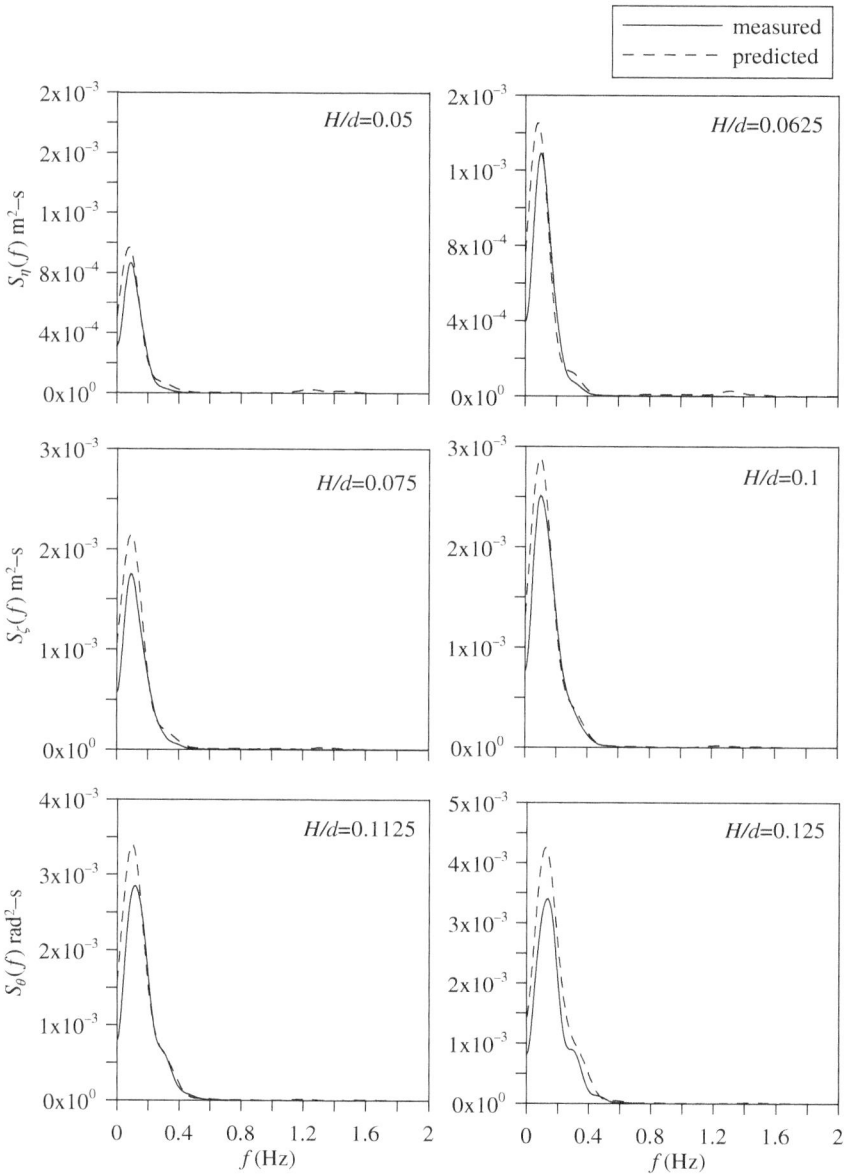

Fig. 7.20   Prediction of solitary wave spectra using HTF.

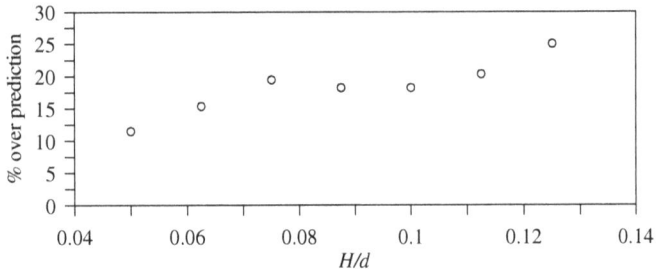

Fig. 7.21   Percentage overestimation.

frequency of the wave elevation. The pitch spectra are found to exhibit a secondary peak at a frequency about 10 times the modal frequency of the wave.

7.5.3.3   *Prediction using TF*

Similar to Stokes and Cnoidal wave tests, the solitary wave spectral densities are predicted from the buoy model response spectra using the HTF. The comparisons of measured and predicted wave elevation spectra for different H/d values are shown in Fig. 7.20. The estimation is always found to be higher than that of the measured particularly around the peak frequency. The variation of the percentage overestimation in estimating solitary wave spectra from HTF is found to increase with an increase in the H/d as can be seen in Fig. 7.21, the results of which are obtained from $H/d = 0.05$–0.125.

7.6   Summary

The critical experimental investigations conducted over a scaled down NDBP buoy presented with the salient results of phase-time variations over the spectral densities aided to effectively identify a characteristic tsunami wave passing. Thus, an existing data collection system deployed in deep sea can act as a tsunami detection sensor and issue early warning about an approaching tsunami well before it reaches the shore.

References

Balaji, R., Sannasiraj, S. A., and Sundar, V. (2006). "Tsunami wave interaction with data buoys." *Marine Geodesy*, 29(4), 235–251.

Balaji, R. (2007). "Characterization of sea states from dynamics of discus data buoys." Doctoral thesis, Indian Institute of Technology Madras, India.

Gonzalez, F. I., Milburn, H. M., Bernard, E. N., and Newman, J. C.(1998). "Deep-ocean assessment and reporting of tsunamis (DART®): Brief overview and status report." Proceedings of the International Workshop on Tsunami Disaster Mitigation, 19–22 January 1998, Tokyo, Japan.

Green, D. (2006). "Transitioning NOAA moored buoy systems from research to operations." Proceedings of OCEANS'06 MTS/IEEE Conference, 18–21 September 2006, Boston, MA, CD-ROM.

Meinig, C., Stalin, S. E., Nakamura, A. I., and Milburn, H. B. (2005). "Real-time deep-ocean tsunami measuring, monitoring, and reporting system: The NOAA DART II description and disclosure." NOAA, Pacific Marine Environmental Laboratory (PMEL), Seattle, WA, USA, 1–15.

Milburn, H. B., Nakamura, A. I., and Gonzalez, F. I. (1996). "Real-time tsunami reporting from the deep ocean." Proceedings of the Oceans 96 MTS/IEEE Conference, 23–26 September 1996, Fort Lauderdale, FL, 390–394.

National Research Council (2011). "Tsunami warning and preparedness: An assessment of the U.S. tsunami program and the nation's preparedness efforts." The National Academies Press, Washington, DC.

The National Academy of Sciences Engineering and Medicine (2011). "Tsunami Warning and Preparedness: An assessment of the U.S tsunami program and the Nation's Preparedness efforts," The National Academic Press.

Weinstein, S. A., and Lundgren, P. R. (2008). "Finite fault modeling in a Tsunami warning center context." Pure and Applied Geophysics, 165, 451.

Chapter 8

Effectiveness of Coastal Vegetation on Impacts due to Tsunami

8.1 General

A great deal of attention is being given by researchers worldwide to the mitigation measures for minimizing the impacts of coastal hazards like storm surges and tsunamis along the coastal region. However, any mitigation measure should preferably be eco-friendly and certainly not cause further adverse effects to the region. Traditionally, hard measures such as construction of seawalls, groyne fields, offshore detached breakwaters or a combination of these structures have been employed as measures to protect a coast from natural hazards as well as from perennial erosion. However, these structures could cause environmental concerns and are often quite expensive, requiring regular maintenance for their sustainability. The effect of vegetation in reducing the inundation distance and height was greatly revealed after the Indian Ocean tsunami of 2004. The vegetation at several locations along the affected coast acted as buffers and contributed to the safety of structures along the shore.

8.2 Vegetal Resistance

8.2.1 *General*

The wave run-up characteristics along beaches and marine structures are essential for planners and designers to decide on whether coastal structures at any site have to be of non-overtopping or overtopping type which mainly depends on the purpose for which it is proposed. At a few locations, a certain degree of overtopping is acceptable, whereas in densely populated locations important infrastructures like power plants are not allowed.

In most of the studies carried out in the past, the vegetation was simulated by a group of cylinders of constant height and diameter for regular configuration (Meijer and Van Velzen, 1999; Nepf, 1999). Dudley *et al.* (1998) reported that the flow resistance depends on the density of the vegetation within the group defined as the frontal area of submerged vegetation projected onto a plane perpendicular to the direction of flow per unit volume of flow as well as bending stiffness of the species. Armanini and Righetti (2002) proposed a formula through an analytical two-layer model for the evaluation of vegetation resistance with sparsely distributed bushes under uniform flow conditions.

The interaction of long waves with vegetation has been gaining importance since the great Indian tsunami of 2004, the significant contributions of which are briefly discussed in the subsequent paragraphs.

Furukawa *et al.* (1997) demonstrated through physical model tests that the wave dissipation depends on the forest density and the diameter of the tree trunks. Hamzah *et al.* (1999), Harada and Imamura (2001) and Tanaka *et al.* (2006) have emphasized the attenuation of incident wave energy due to the presence of coastal forests. Mendez and Losada (2004) proposed an empirical model for the estimation of transmission coefficient for a vegetation patch. Through, physical modelling Hiraishi and Harada (2003) proved that the reduction in tsunami heights increases with an increase in density of the green belt. Harada and Imamura (2005) proposed a relation between the forest density and the diameter of trunks and the effects of reduction of the inundation heights and distance due to tsunamis. Yao *et al.* (2015) have reported that the normalized wave run-up over the beach decreases with an increase in both normalized incident wave height and forest density, and an empirical formula with the density incorporated was proposed. Their study also highlighted the importance of tree distribution to wave interaction with vegetation on the slope when the forest density was unaltered, and that run-up reduction difference between tandem and staggered arrangements of the trees could reach up to 20%. Hu *et al.* (2014) have proposed an empirical relation between drag coefficient and Reynolds number for combined wave and current flow past vegetation, which can be useful for numerical modelling. The review of the literature has shown that while considerable work has been carried out in the past on the pressures on walls due to random waves, not much information is available on the pressures on walls fronted by vegetation, in particular due to long waves.

8.2.2   *Submerged vegetation*

In general, there are two types of vegetation, *viz.*, rigid (normally woody or arborescent plants) and flexible (herbaceous plants). A simplified model for evaluating the resistance with hydraulic and vegetative parameters for emerged and submerged vegetation was proposed by Wu *et al.* (1999). The experimental results from their studies revealed that the roughness coefficient decreases with an increase in depth under the emerged condition. For fully submerged conditions, the vegetative roughness coefficient was reported to initially increase with depth, at lower depths, and reach an asymptotic constant value as the water level continues to rise. Freeman *et al.* (2000) conducted more than 220 experiments in a flume, using 20 different plant species. The most commonly used equation for flow resistance is Manning's equation.

$$U = n^{-1}R^{2/3}S^{1/2}, \tag{8.1}$$

where U is the average velocity, R is the hydraulic radius, S is the channel slope and n is Manning's resistance coefficient.

In order to deal with the flexible vegetation, a cantilever beam theory was used to compute the deflection of the vegetation. A computer model, complex flow simulations (COMSIM), was developed and applied to study the influence of vegetation on the flow resistance based on the assumption that its main impacts on flow is that it causes a drag, resulting in momentum losses (Fischenich, 2000), which may be due to several factors like stiffness, diameter, height, distribution, density and type of vegetation that influence the resistance.

8.2.3 *Non-submerged vegetation*

In general, non-submerged vegetation dissipates a greater energy and momentum from the flow compared to that due to the presence of submerged vegetation. This type of vegetation is often found to be in a place where the roughness is high. Long trees like coniferous trees alter the friction factors greatly which was validated by Kouwen and Fathi-Moghadam (2000) using coniferous tree saplings in flume experiments and large coniferous trees in air experiments with the mean flow velocity that caused bending of the vegetation thus leading to an increase in the submerged momentum absorbing area. The friction factor for non-submerged roughness relates to the flow velocity and a species-specific vegetation index that accounts for the effects of shape, flexibility and biomass. The vegetation index (E_{vi}) was considered to be one of the important parameters to account for the effects of leaf density, shape and rigidity of individual trees. With its height, mass and fundamental natural frequency, the vegetation index is estimated through the relation given in Eq. (8.2).

$$E_{vi} = f_1^2 \left(\frac{m_s}{l}\right),\qquad(8.2)$$

where the total mass of the tree $m_s = ml$, in which m is mass per unit length, f_1 is the first mode of the natural frequency, l is the length of the tree and E_{vi} is called the "vegetation index" of a tree species.

In the vegetation–flow interaction, the resistance to the fluid flow is considered to be characterized by a number of friction factors. The resistance coefficients that are generally considered are Manning's n, Darcy–Weisbach friction factor f and Chezy's constant C. There are also a few studies in which the vegetal drag has been investigated. The Darcy f has been quantified in the past through measurements in open channel experiments (Unny et al., 1969; Petryk and Bosmajian, 1975; Chen, 1976; Werth, 1997) using various vegetation models. In general, it may be noted that the interaction between vegetation and flow has not been modelled appropriately and the elastic characteristics of vegetation have not been given due consideration. However, the previous investigators reported that quantifying the resistance using f has several advantages such as (i) the flow being taken past enclosed/partially enclosed vegetal volumes, (ii) ease of measurements in the laboratory and (iii) possibility of quantification of wall effects in the open channel, etc.

8.3   Physical Modelling

8.3.1   *General*

The parameters that have been considered herein include the size, elastic modulus of the vegetal stems, configuration of the vegetal patch or green belt, the spacing between the individual plants within the green belt etc. Quantification of friction factor and energy loss in steady uniform flow are of primary interest and have practical relevance when long waves are concerned. For the measurement of pressures and forces on the structures mounted at a distance from the shoreline/reference line and run-up on mild slope of 1:30, a number of wave and beach parameters were considered in the wave flume experiments. The following variables are of interest for both currents and waves for investigating the response function such as resistance, force or run-up.

Vegetal Response
$$= f(\rho, g, h, H, T, B_s, D, D_b, BG, f_1, l, SP, E, L, \beta, R_u, V),\qquad(8.3)$$

Table 8.1. Range of dimensionless parameters covered for the physical model study.

Parameters	Wave flume (Model)	Steady flow (Model)
Keulegan–Carpenter number, $KC = \left(\frac{U_{\max}}{B_s}()\right)$	31–145	Nil
VFP $= \left(\frac{EI(BG/D)}{\rho H l^3 V_{avg}^2(SP/D)}\right)$	0.006–2.31	0.001–14.34
Vegetal parameter $= \left(\frac{BG*SP}{D^2}\right)$	94–8333	94–8333
Reduced velocity, $V_r = \left(\frac{V_{avg}}{f_1 D_t}\right)$	11–166	2.1–83
Surf similarity parameter, $\xi = \left(\frac{\tan\beta}{\sqrt{\frac{H}{L_o}}}\right)$	0.17–0.42	Nil
Relative run-up height $= \left(\frac{R_u}{H}\right)$	0.5–1.01	Nil
Froude number, $F_r = \left(\frac{V_{avg}}{\sqrt{gh_{avg}}}\right)$	1.19–1.51	0.32–0.80

where ρ is mass density of water; g is gravitational acceleration; h is depth of water in the flume; H is wave height; T is wave period; B_s is width of the structure; D is diameter of the vegetation; D_b is diameter at the root of the vegetation; BG is width of green belt; f_1 is frequency of first mode of the vegetal stem; l is height of the vegetation; SP is centre–to–centre spacing between vegetation; E is modulus of plant stiffness; L is wave length; β is beach slope; R_u is run-up; V is flow velocity. The h refers to h_s which is depth of water at the toe of the structure; h_{avg} is average depth of water upstream and downstream of flow in the open channel.

The variables seen in Eq. (8.3) are grouped as per Buckingham's Pi theorem. The parameters thus obtained are tabulated in Table 8.1 in which $U_{\max}$ is the flow velocity (V) and L_o is the deepwater wave length.

Among all the aforesaid parameters, the vegetation-flow parameter (VFP) is newly introduced to obtain a unique parameter that combines the effects of structural rigidity of the individual model stem, flow and vegetation parameters. It is prudent to now recall the fact that the earlier investigators have used the Vegetation Index as a parameter to classify vegetation (Fathi-Moghadam, 2007). However, it is to be noted that the Vegetation Index does not include parameters relating to rigidity EI, vegetation parameters BG, SP/D and the flow parameters. Hence, it is proposed that the present parameters derived from this work will be more relevant and comprehensive in view of the hydroelastic processes being considered.

8.3.2 *Laboratory study*

Quantification of friction factor and energy loss past vegetation is of primary interest and has practical relevance. However, while determining the forces on the structures mounted at a distance from the shoreline/reference line, with the presence and absence of vegetation, a number of wave and beach parameters should be considered. On the other hand, the vegetation on the seaside of existing structures during the ingress of the great Indian Ocean tsunami have acted as buffers in reducing the inundation heights and distance into the land, the phenomena of which is not well understood. Hence, for the purpose of studying the different factors, a series of experiments had been carried out under different laboratory conditions for specific cases (Noarayanan, 2009).

8.3.3 *Modelling the green belt*

In order to understand the hydro-elastic interaction of the flow with vegetal stems, a suitable material for model had to be identified to represent coastal vegetation in the real world. One of the guiding parameters for this purpose is Young's modulus, E. The common timber would have an E value in the range from 10.05 GPa to 15 GPa. The mangrove has one of the highest E values of 20.03 GPa. In order to cover a wide range of E, a reference value of 14 GPa has been assumed for the field which had to be modelled for laboratory tests. In order to scale down the prototype values and identify the model material, assuming that the flow is due to storm surge or tsunami, and also considering the experimental ranges of flow (including velocities and flow depth), a guiding scale ratio of 1:40 was adopted. This would mean that the E value for the model material should be about 0.35 GPa, which is quite difficult to identify.

The most practical option for achieving the aforementioned criteria is to consider EI as a single parameter and scaling the rigidity. In laboratory studies, the material considered was *Poly Ethylene* with an E value of about 3.8 GPa which is about 10 times that of the required value. Since the rigidity modelled herein is higher than the said variation, it could be compensated with the variation in moment of inertia, i.e., I of the material. Having chosen the model material, the typical prototype dimensions of vegetal stems are fixed in the range of 100–400 mm. Accordingly, using the Froude model law of scaling, the rigidity with the scale factor of [SF]5 — the base diameter for model vegetation — will fall in the range of 1.65–5.5 mm. The base diameter as derived earlier only ensures that the bending action of the vegetal stems is properly scaled. Thus, the advantage of modelling the rigidity rather than Young's modulus is clearly brought out. However, for accurate hydrodynamic interaction of the flow and stems, the hydrodynamic forces need to be correctly scaled as per Froude's Law. Considering the drag/inertia force regime (Chakrabarti, 1983) the corresponding diameters for the exposed part of vegetal stems were arrived.

8.4 Role of Vegetation on Wave-Induced Pressures on Wall

Pressures on any coastal structure are of engineering interest, while application of bio-shield or green or vegetation patch is also considered. The dynamic pressures exerted on a vertical wall is due to the action of Cnoidal waves that have been measured in the presence and absence of model vegetation in front of the wall. For the purpose of identifying the role of vegetation on wave-induced pressure on a wall, the studies were carried out by varying the vegetal configuration, *viz.*, regular and staggered placements, and comparing with the results in the absence of vegetation. The measurements along the depth provided significant details on how the dynamic pressure varied with the waves for different vegetal configurations.

If one needs to model this kind of physics in laboratory, i.e., to study the effect of vegetation on flow attenuation during the aforesaid extreme events, the flow and vegetal parameters used should be in line with realistic ranges. The behaviour of steady flow past vegetation was carried out in an open channel. The parameters that have been considered herein include the size, elastic modulus of the vegetal stems, configuration of the vegetal patch or green belt, as well as, the spacing between the individual plants within the green belt, etc. A typical setup for the measurement of forces on the structure in absence of vegetation and with configuration of vegetation is shown in Fig. 8.1.

Herein, G is the spacing between the vegetation and the structure and B is the width of the structure. The solution for forces on simple structures like circular as well as non-circular cylinders is well entrenched in the literature. While commencing the studies, initial tests with the model structure mentioned earlier instrumented with a load cell for measuring the in-line forces and rigidly fixed on the sloping bed without the vegetation was subjected to waves of different frequencies and heights. The measured forces compared with the solution of Isaacson (1979) in

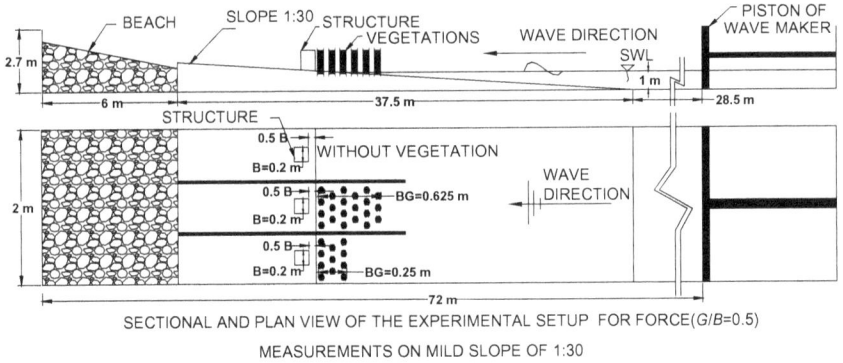

SECTIONAL AND PLAN VIEW OF THE EXPERIMENTAL SETUP FOR FORCE(G/B=0.5)

MEASUREMENTS ON MILD SLOPE OF 1:30

Fig. 8.1 Experimental setup for force measurements in the wave flume for $G/B = 0.5$.

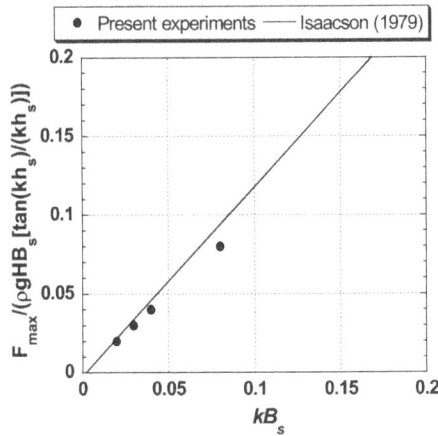

Fig. 8.2 Comparison of present experimental results with Isaacson (1979).

Fig. 8.2 exhibit a reasonable agreement, thus validating the measurement of forces.

The forces exerted on the model structure due to Cnoidal waves of moderate to high Ursell parameter, $U_r = HL^2/h^3$, i.e., from 18 to 700 have been measured in the absence and presence of vegetation on its seaside. In order to make the presentation of results meaningful and permitting a direct application of model results to field, the force is expressed in a dimensionless form as $F^* = [F_{max}/(0.5\rho g H^2 B_s)]$. The variation of dimensionless force with reduced velocity V_r is investigated. The said results have been reported by Sundar *et al.* (2011) and Noarayanan *et al.* (2012a).

The KC number is usually defined as $(U_{max}T)/D$, wherein U_{max} is approximated as $c = \sqrt{[gh_s(1+(H/h_s))]}$ (Wiegel, 1964) and $D = B_s$ which is the breadth of the structure normal to the wave direction. As stated earlier, the tests were carried out with Cnoidal waves of four different periods, the results of which are discussed in the subsequent paragraphs.

The variation of F^* as a function of V_r, for the four different G/B for the range of lower V_r of 11–14 and for BG/SP ranging between 3 and 8, are projected in Fig. 8.3 and that for $BG/SP = 13$–17 and 26 are shown in Figs. 8.4 and 8.5, respectively. Herein, the plots are combined irrespective of h_s/gT^2. Once again, the discussion presented earlier for the other V_r range is valid here. The results demonstrate that the F^* significantly reduces

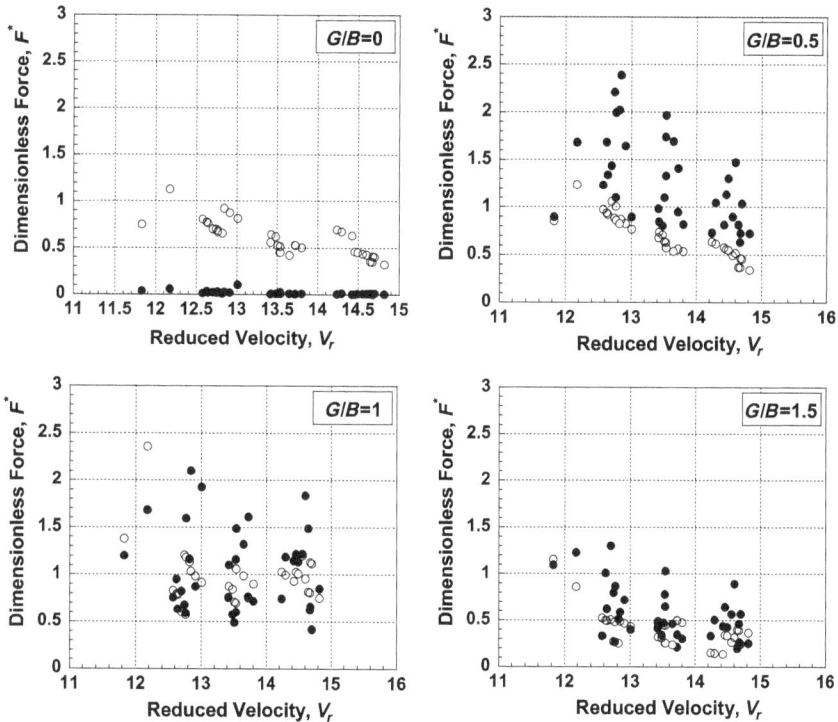

Fig. 8.3   Variation of dimensionless force with reduced velocity for $V_r = 11$–14 and $BG/SP = 3$–8 (∘ is No Veg, • is Veg).

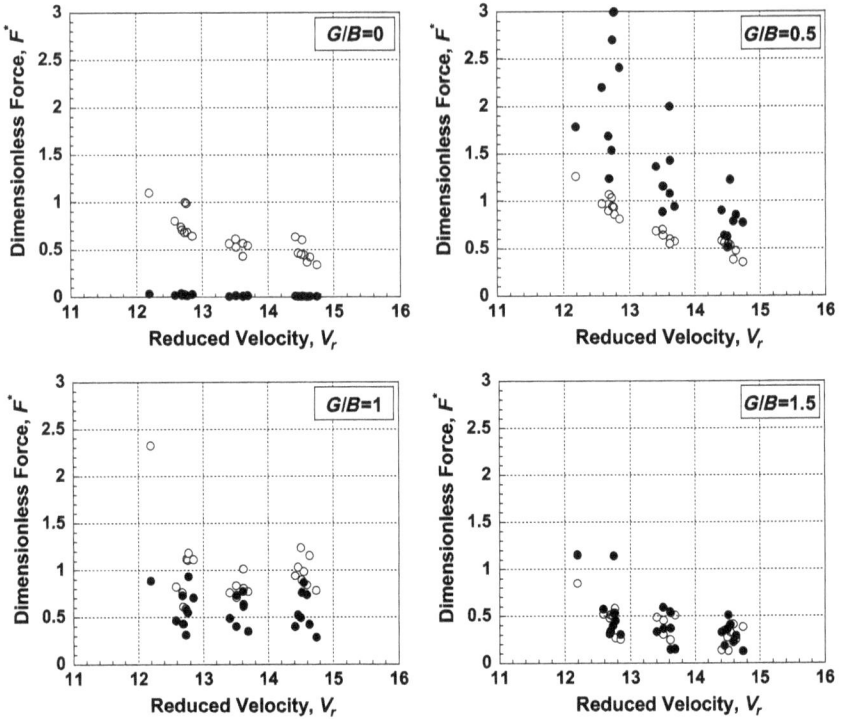

Fig. 8.4   Variation of dimensionless force with reduced velocity for $V_r = 11$–14 and $BG/SP = 13$–17 (○ is No Veg, ● is Veg).

in the presence of vegetation adjoining the structure and as the distance between the vegetation and the structure increases the F^* for the structure in the presence of vegetation initially increases, for G/B varied from 0 to 0.5. With a further increase in G/B, the rate of increase reduces drastically and when G/B is at 1.5, the forces are almost same or even lesser for the structure fronted with vegetation. This suggests as mentioned earlier that for the coastal vegetation to be effective in reducing the forces on structures along shore, it should be at least twice the width of the structure (distance measured normal to shoreline) or more. The forces in general increases by about 100% with the gap ratio G/B of 0.5 and this increase is about –30 to 30% in case of $G/B = 1.0$ and –50 to 20% in case of $G/B = 1.5$. It is expected that the force will reduce when $G/B > 1.5$. Similar results for the range of larger V_r of 130–166 and for $BG/SP = 3$–8, 13–17 and 26 are shown in Figs. 8.6, 8.7 and 8.8, respectively.

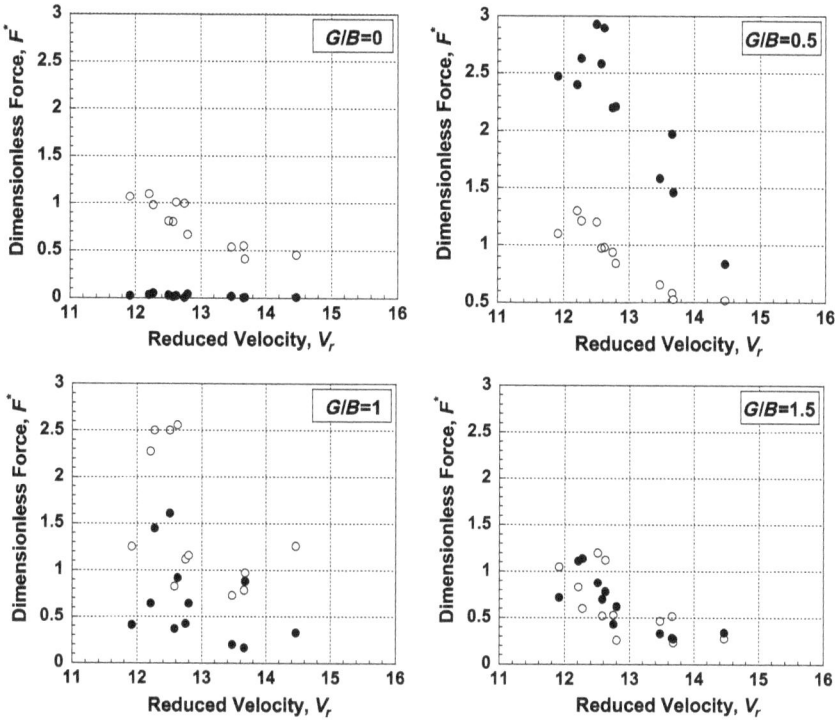

Fig. 8.5   Variation of dimensionless force with reduced velocity for $V_r = 11$ to 14 and $BG/SP = 26$ ($\circ$ is No Veg, $\bullet$ is Veg).

8.5   Friction Factors

8.5.1   *General*

The percentage of reduction in the hydrodynamic energy due to vegetation is also governed by the diameter of the average member within the vegetation, its height, distribution, density and its elastic properties. The study aims to investigate the Darcy–Weisbach and Manning's friction coefficient n friction factors for various flow as well as vegetative parameters considering the aforementioned green belt effects. Manning's n and Darcy–Weisbach's f friction factors for various flow as well as vegetative parameters were determined for regular and staggered configurations of the vegetation from the tests. New empirical equations are proposed for estimating Manning's friction coefficient n and Darcy–Weisbach's friction factor f, independent of vegetal submergence and undeflected plant height.

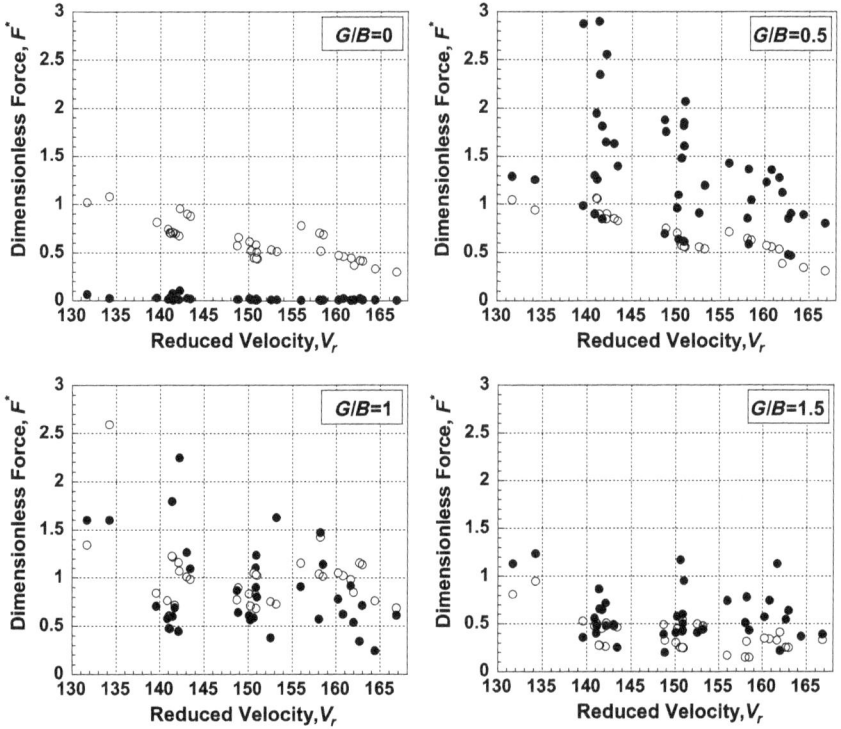

Fig. 8.6   Variation of dimensionless force with reduced velocity for $V_r = 130\text{--}166$ and $BG/SP = 3\text{--}8$ ($\circ$ is No Veg, $\bullet$ is Veg).

8.5.2   *Darcy–Weisbach friction factor f*

Since the trees could be characterized similar to cantilever beams, the mechanical characteristics may be brought in through the resonance frequency of the first mode of vibration, namely the natural frequency of the vegetal stems. The natural frequency could be related to the "Vegetation Index" as defined by Fathi-Moghadam (2007). The resonance frequencies f_j (with $j = 1, 2, 3, \ldots, n$, where f_1 is the fundamental or base natural frequency and $f_2, \ldots, n$ are higher modes of natural frequencies) of a linear and homogeneous beam depend upon its length l, mass per unit length m, second moment of inertia I, modulus of elasticity E, as well as a dimensionless parameter λj which in turn is a function of beam geometry and the boundary conditions under which it is tested. The relationship between the

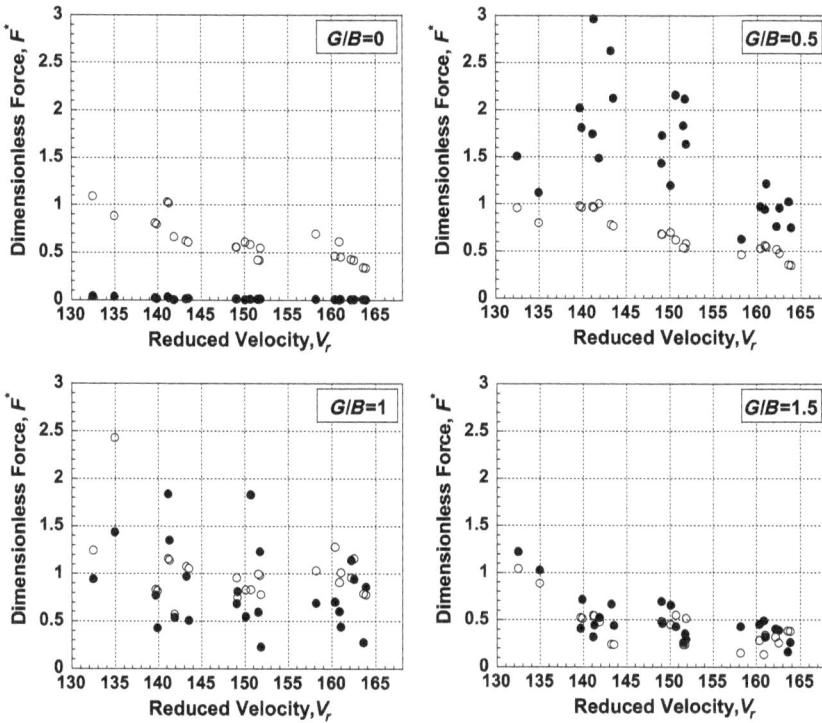

Fig. 8.7   Variation of dimensionless force with reduced velocity for $V_r = 130$–166 and $BG/SP = 13$–17 (○ is No Veg, ● is Veg).

resonance frequency and the aforesaid variables is given by the following equation by Timoshenko and Gere (1961).

$$f_j = \frac{\lambda_j^2}{2\pi} \left(\frac{EI}{ml^4} \right)^{0.5}. \tag{8.4}$$

The parameters in Eq. (8.4) characterize the height, mass or leaf density and the moment of inertia of a tree. Herein, the beam length l is taken to be the height of the vegetal stem. According to Fathi-Moghadam (2007), on transferring the measurable parameters to the right side, equation for the first mode of the natural frequency f_1 will be as follows:

$$E_{vi} = f_1^2 \left(\frac{m_s}{l} \right), \tag{8.5}$$

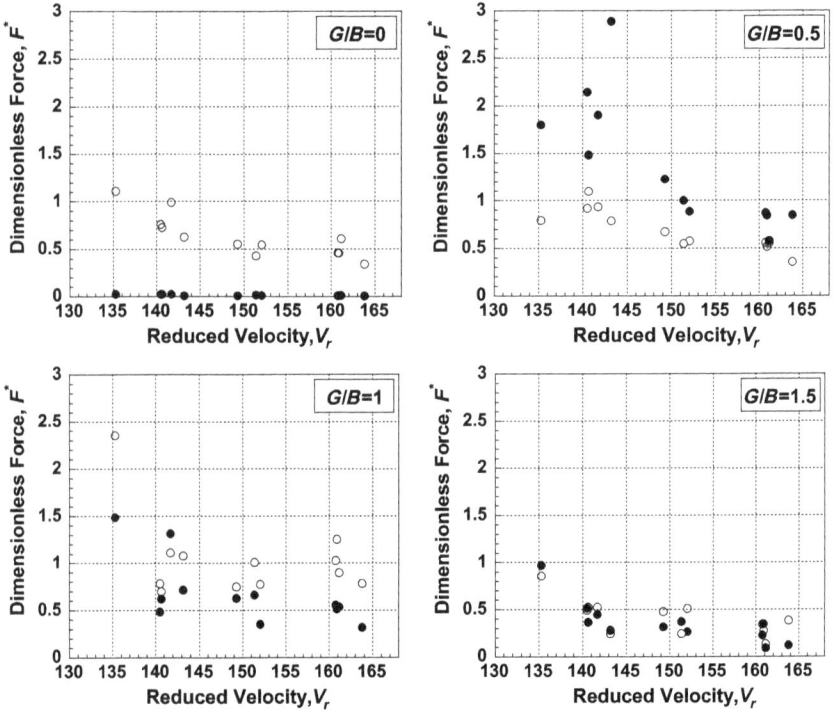

Fig. 8.8   Variation of dimensionless force with reduced velocity for V_r =130–166 and $BG/SP = 26$ (○ is No Veg, ● is Veg.)

where the total mass of the tree $m_s = ml$, and E_{vi} is called the "vegetation index" of a tree species. By measuring the height, mass and fundamental natural frequency of the tree, the vegetation index can be estimated through Eq. (8.5).

8.5.2.1   Determination of head loss due to vegetation

The f Weisbach friction factor is given by the following:

$$f_h = \frac{8H_f gh}{LV^2} \tag{8.6}$$

where H_f is head loss due to friction; L is distance between the sections. Utilizing the Bernoulli equation, the energy loss of head due to friction, H_f can be determined as

$$\frac{V_u^2}{2g} + h_u = \frac{V_d^2}{2g} + h_d + H_f. \tag{8.7}$$

In the aforementioned equations, V, h and g were defined earlier and the suffixes u and d correspond to the respective variables on upstream and downstream of flow, respectively. However, H_f as seen in Eq. (8.7) also includes the contribution of flume walls and the bed to the flow resistance. The energy loss only due to vegetation $H_{f(\text{veg})}$ may be calculated by deducting the energy losses due to side and bottom walls $H_{f(\text{wall})}$, from the total energy losses $H_{f(\text{veg}+\text{wall})}$.

$$H_{f(\text{veg})} = H_{f(\text{veg}+\text{wall})} - H_{f(\text{wall})}, \qquad (8.8)$$

where

$$H_{f(\text{veg}+\text{wall})}$$
$$= \left\{ \left(\frac{V_{u(\text{veg}+\text{wall})}^2 - V_{d(\text{veg}+\text{wall})}^2}{2g} \right) + (h_{u(\text{veg}+\text{wall})} - h_{d(\text{veg}+\text{wall})}) \right\},$$
$$\qquad (8.9)$$

$$H_{f(\text{wall})} = \left\{ \left(\frac{V_{u(\text{wall})}^2 - V_{d(\text{wall})}^2}{2g} \right) + (h_{u(\text{wall})} - h_{d(\text{wall})}) \right\} \qquad (8.10)$$

and

$$H_{f(\text{veg})} = \left\{ \left(\frac{V_{u(\text{veg}+\text{wall})}^2 - V_{d(\text{veg}+\text{wall})}^2}{2g} \right) + (h_{u(\text{veg}+\text{wall})} - h_{d(\text{veg}+\text{wall})}) \right\}$$
$$- \left\{ \left(\frac{V_{u(\text{wall})}^2 - V_{d(\text{wall})}^2}{2g} \right) + (h_{u(\text{wall})} - h_{d(\text{wall})}) \right\}. \qquad (8.11)$$

Herein, $V_{u(\text{veg}+\text{wall})}$ is the upstream velocity measured in the presence of vegetation, $V_{u(\text{wall})}$ is the upstream velocity measured in the absence of vegetation; $h_{u(\text{veg}+\text{wall})}$ is the depth of flow upstream of the vegetation; $h_{u(\text{wall})}$ is the depth of flow at the upstream location in the absence of the vegetation; $V_{d(\text{veg}+\text{wall})}$ is the velocity measured with vegetation on the downstream; $V_{d(\text{wall})}$ is the velocity measured without vegetation on the downstream; $h_{d(\text{veg}+\text{wall})}$ is the depth of flow downstream of the vegetation; $h_{d(\text{wall})}$ is the depth of flow downstream in the absence of the vegetation; $V_{d(\text{veg}+\text{wall})}$ is the velocity measured with vegetation downstream. It is to be noted that the aforesaid approach consistently eliminates the presence of wall effects in the measurements, irrespective of the location of upstream and downstream measurement points.

8.5.3   *Manning's co-efficient n*

8.5.3.1   *Determination of Manning's "n" due to vegetation*

Manning's n is traditionally defined as follows:

$$n = \frac{1}{V} R_h^{\left(\frac{2}{3}\right)} S^{\left(\frac{1}{2}\right)}, \tag{8.12}$$

where R_h is the hydraulic radius (flow area/wetted perimeter); S is bed or energy slope (dimensionless); V is the flow velocity; n is Manning's n.

As mentioned earlier, utilizing the Bernoulli equation, the energy loss of head due to friction, H_f can be determined as the following:

$$\frac{V_u^2}{2g} + h_u = \frac{V_d^2}{2g} + h_d + H_f. \tag{8.13}$$

In the aforementioned equations, V, h and g were defined earlier and the suffixes u and d correspond to the respective variables on upstream and downstream of flow, respectively. H_f is the head loss due to friction. In order to consider the wall effects

$$n_{(\text{veg})} = n_{(\text{veg+wall})} - n_{(\text{wall})}, \tag{8.14}$$

where

$$n_{(\text{veg+wall})} = \frac{1}{V} R_{h(\text{veg+wall})}^{\left(\frac{2}{3}\right)} S_{(\text{veg+wall})}^{\left(\frac{1}{2}\right)}. \tag{8.15}$$

$$n_{(\text{wall})} = \frac{1}{V} R_{h(\text{wall})}^{\left(\frac{2}{3}\right)} S_{(\text{wall})}^{\left(\frac{1}{2}\right)} \tag{8.16}$$

and

$$R = \frac{\text{depth of flow}}{\text{wetted perimeter}}. \tag{8.17}$$

Now from the Bernoulli's equation,

$$H_{f(\text{veg+wall})}$$
$$= \left\{ \left(\frac{V_{u(\text{veg+wall})}^2 - V_{d(\text{veg+wall})}^2}{2g} \right) + \left(h_{u(\text{veg+wall})} - h_{d(\text{veg+wall})} \right) \right\}. \tag{8.18}$$

Now, the energy slope in terms of H_f and the distance over which it occurs, BG, may be defined as follows:

$$S_{(\text{veg+wall})} = \frac{H_{f(\text{veg+wall})}}{BG}. \tag{8.19}$$

However, using the measurements without the vegetal patch

$$H_{f(\text{wall})} = \left\{ \left(\frac{V_{u(\text{wall})}^2 - V_{d(\text{wall})}^2}{2g} \right) + (h_{u(\text{wall})} - h_{d(\text{wall})}) \right\} \qquad (8.20)$$

and

$$S_{(\text{wall})} = \frac{H_{f(\text{wall})}}{BG}. \qquad (8.21)$$

Now, $n_{(\text{wall})}$ may be estimated using Eq. (8.16). Defining the $n_{(\text{veg})}$ as Eq. (8.22):

$$n_{(\text{veg})} = \left\{ \left(\frac{1}{V_{(\text{veg}+\text{wall})}} \right) * (R)_{h(\text{veg}+\text{wall})}^{(2/3)} * S_{(\text{veg}+\text{wall})}^{(1/2)} \right\}$$
$$- \left\{ \left(\frac{1}{V_{(\text{wall})}} \right) * (R)_{h(\text{wall})}^{(2/3)} * S_{(\text{wall})}^{(1/2)} \right\}, \qquad (8.22)$$

where calculation of $R_{h(\text{veg}+\text{wall})}$ after considering the effects of presence of vegetation is given as follows:

$$R_{h(\text{veg}+\text{wall})} = \left\{ \frac{\left(\frac{(B_c * h_{\text{avg}(\text{veg}+\text{wall})}) - (nD * h_{\text{avg}(\text{veg}+\text{wall})})}{2h_{\text{avg}(\text{veg}+\text{wall})} + [B_c - ((B_c/SP) * D)] + 2(B_c/SP) * h_{\text{avg}(\text{veg}+\text{wall})}} \right) + \left(\frac{B_c * h_{\text{avg}(\text{wall})}}{B_c + 2 * h_{\text{avg}(\text{wall})}} \right)}{2} \right\}. \qquad (8.23)$$

$$R_{h(\text{wall})} = \left(\frac{B_c * h_{\text{avg}(\text{wall})}}{B_c + 2 * h_{\text{avg}(\text{wall})}} \right). \qquad (8.24)$$

Herein, $V_{u(\text{veg}+\text{wall})}$ is the upstream velocity measured in the presence of vegetation, $V_{u(\text{wall})}$ is the upstream velocity measured in the absence of vegetation; $h_{u(\text{veg}+\text{wall})}$ is the depth of flow upstream of the vegetation; $h_{u(\text{wall})}$ is the depth of flow at the upstream location in the absence of the vegetation; $V_{d(\text{veg}+\text{wall})}$ is the velocity measured with vegetation on the downstream; $V_{d(\text{wall})}$ is the velocity measured without vegetation on the downstream; $h_{d(\text{veg}+\text{wall})}$ is the depth of flow downstream of the vegetation; $h_{d(\text{wall})}$ is the depth of flow downstream in the absence of the vegetation; $V_{d(\text{veg}+\text{wall})}$ is the velocity measured with vegetation on the downstream; $R_{h(\text{veg}+\text{wall})}$ is the average hydraulic radius at two sections; one is exactly at vegetation cross-section and the second is the hydraulic radius taken in between the vegetal stems in the flow direction; $R_{h(\text{wall})}$ is the average hydraulic radius in the absence of vegetation; $S_{(veg+wall)}$ is the energy slope with vegetation and $S_{(wall)}$ is the energy slope without vegetation; B_c is the width of the channel; $n_{(veg)}$ is Manning's n only due to vegetation; $n_{(veg+wall)}$ is Manning's n due to vegetation and side walls; $n_{(wall)}$ is Manning's n only due to walls. It is to be noted that the aforesaid approach consistently eliminates the presence of wall effects in the measurements, irrespective of the location of upstream and downstream measurement points.

Manning's n has been computed from the experimental measurements as outlined earlier. Manning's n is studied as a function of various parameters as identified in the previous section (Noarayanan *et al.*, 2013) and the same is reported and discussed in this section

8.6   Relationship of Energy Loss with VFP

The f and n are important parameters for vegetal drag or head loss. However, its direct use in the design of Greenbelt is not quite straightforward. One needs to have a fair understanding of the energy loss and its direct relation to the vegetal and flow parameters in order to perform a design calculation for green belt. In this study, the normalized Energy Loss $E_{L(\text{Reg})}$ and $E_{L(\text{ZZ})}$, with reference to the upstream energy content is considered as against the VFP so that a direct relation between the green belt parameters and drop in energy within the green belt may be obtained. The variation of $E_{L(\text{Reg})}$ as a function of VFP is displayed in Figs. 8.9(a) and 8.9(b) and the variation of $E_{L(\text{ZZ})}$ as a function of VFP is displayed in Figs. 8.10(a) and 8.10(b) Since the f and n are directly related, the plot looks similar to that of f and n as presented previously. The $E_{L(\text{Reg})}$ is found to increase from about 0.1 at VFP $\sim$0.001 to 0.8 at VFP 1.5. It has to be recalled that the maximum$f$ and n have been observed at the same value of VFP as reported in earlier figures. Beyond VFP $= 1.5$, the $E_L(\text{Reg})$ reduces and is about 0.30 at VFP $\sim$15. Furthermore, the effects of wake and proximity interactions are once again brought out in the$E_{L(\text{Reg})}$ also as the variation of $E_{L(\text{Reg})}$ for $SP/D = 3.75$ show a clearly superior performance than the other combinations. Similar trend is observed in the Energy Loss with VFP for the vegetation in the staggered configuration. The energy loss is about 150–200% more (0.4–0.9) for $SP/D = 3.75$ while the range of $E_{L(\text{Reg})}$ for higher SP/D is between 0.05–0.4. The mechanisms of wake and proximity effects are discussed earlier and hence are not repeated here.

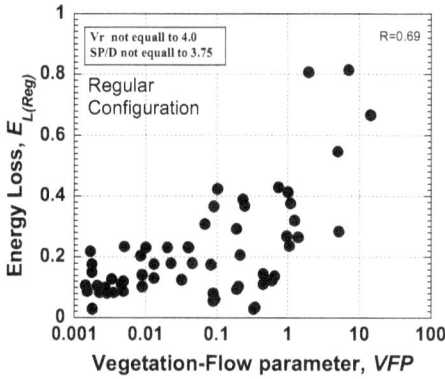

(a)

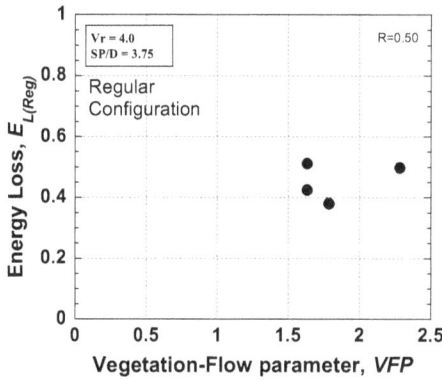

(b)

Fig. 8.9   (a) Variation of energy loss with relative rigidity for various SP/D in regular configuration. (b) Variation of energy loss with relative rigidity for various SP/D in regular configuration.

An attractive proposition is to obtain a direct relationship between $E_{L(\text{Reg})}$ and the independent parameters concerning the flow and green belt. Once again, the multiple regression procedure is extended for $E_{L(\text{Reg})}$ and other parameters. However, in this case, to keep the relationship simple only F_r and VFP are used in the regression procedure.

While $SP/D \neq {\sim}3.75$ and $V_r \neq {\sim}4.0$.

$$E_{L(\text{Reg})} = 0.4369 * \left(\frac{EI * (BG/D)}{\rho h_u V_{\text{avg}}^2 l^3 (SP/D)} \right)^{0.2038} \left(\frac{V_{\text{avg}}}{\sqrt{gh_{\text{avg}}}} \right)^{0.5413}, \quad (8.25)$$

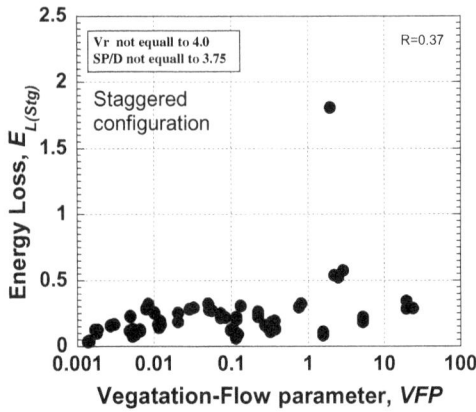

(a)

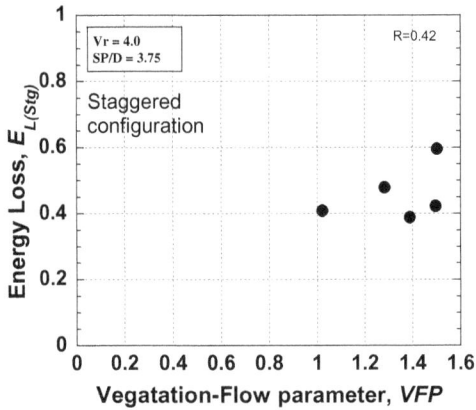

(b)

Fig. 8.10 (a) Variation of energy loss with relative rigidity for various SP/D in staggered configuration. (b) Variation of energy loss with relative rigidity for various SP/D in staggered configuration.

while $SP/D = {\sim}3.75$ and $V_r = {\sim}4.0$.

$$E_{L(\mathrm{Reg})} = 4.25 * \left(\frac{EI * (BG/D)}{\rho h_u V_{\mathrm{avg}}^2 l^3 (SP/D)} \right)^{0.2357} \left(\frac{V_{\mathrm{avg}}}{\sqrt{gh_{\mathrm{avg}}}} \right)^{3.0015}. \quad (8.26)$$

It should be noted that the regression coefficient for Eq. (8.25) is lesser than Eq. (8.26). However, this relationship directly relates $E_{L(\mathrm{ZZ})}$ and VFP for a given F_r and hence is more attractive and simpler to use.

Similar relationships for the staggered configuration are presented in the following equations to obtain a direct relationship between $E_{L(ZZ)}$ and the independent parameters concerning the flow and green belt. Once again, the multiple regression procedure is extended for $E_{L(ZZ)}$ and other parameters. However, in this case, to keep the relationship simple only F_r and VFP are used in the regression procedure. This gives the following relationship for $E_{L(ZZ)}$ with a regression coefficient of 0.75 and 0.9.

While $SP/D \neq \sim3.75$ and $V_r \neq \sim4.0$

$$E_{L(zz)} = 0.171 * \left(\frac{EI * (BG/D)}{\rho h_u V_{avg}^2 l^3 (SP/D)} \right)^{0.1028} \left(\frac{V_{avg}}{\sqrt{gh_{avg}}} \right)^{-0.3949}, \quad (8.27)$$

while $SP/D = \sim3.75$ and $V_r = \sim4.0$

$$E_{L(zz)} = 0.099 * \left(\frac{EI * (BG/D)}{\rho h_u V_{avg}^2 l^3 (SP/D)} \right)^{1.2722} \left(\frac{V_{avg}}{\sqrt{gh_{avg}}} \right)^{-1.7077}. \quad (8.28)$$

It should be noted that the regression coefficient for Eq. (8.27) is lesser than Eq. (8.28). However, this relationship directly relates $E_L(ZZ)$ and VFP for a given F_r and hence is more attractive and simpler to use.

8.7   Role of Vegetation on Beach Run-Up

The run-up due to ocean waves on an open coast and over coastal structures is an essential parameter in their planning, design and its effectiveness in protecting the land against natural hazards. Although several formulae have been derived in the past for the estimation of run-up through laboratory studies for a variety of wave front structures, there still remain certain uncertainties pertaining to flat slopes and long waves. Hence, the study on the effect of vegetation and its influence on wave run-up has been taken up along with pressures and forces on modelled structures as described earlier. It was well evidenced that the presence of vegetation could effectively reduce the wave run-up over the slope by about 50%. Recent study by Yao (2018) on reducing the wave run-up by emergent vegetation provides essential information on the modelling aspects of vegetation based on Boussinesq's equation of solitary wave. Yao (2018) stated that both the wave run-up and drag coefficient can be decreased by increasing the forest density. The effect of vegetation on the run-up due to regular and Cnoidal waves have been discussed by Noarayanan et al. (2012b).

A comprehensive experimental investigation to study the effect of vegetation in attenuating the run-up over a plane slope of 1:30 due to the action

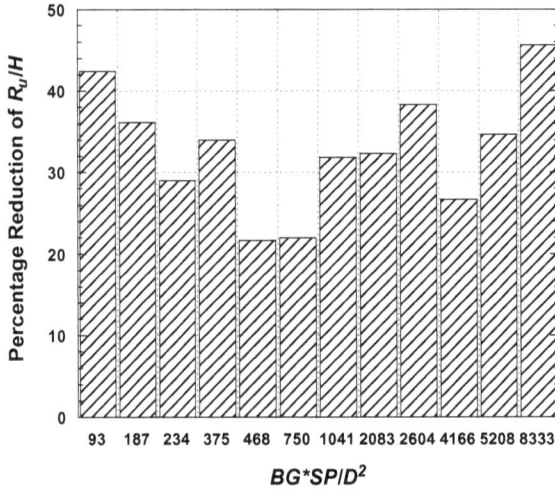

Fig. 8.11   Variation of percentage reduction of R_u/H with BG^*SP/D^2 — random waves.

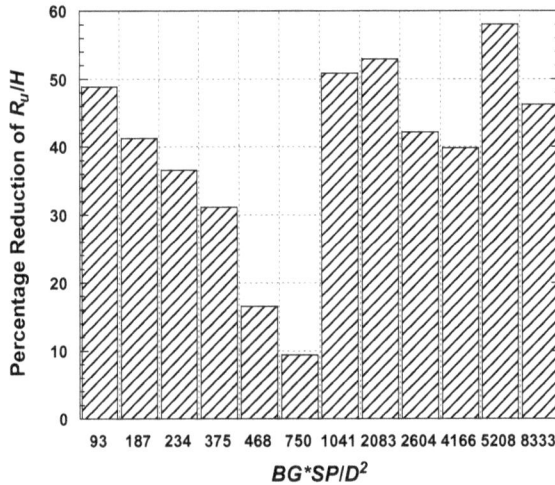

Fig. 8.12   Variation of percentage reduction of R_u/H with BG^*SP/D^2 — Cnoidal waves.

of regular/Cnoidal/solitary waves as well as random waves of predefined spectral characteristics has been carried out. The effectiveness of vegetation of different characteristics in reducing the run-up due to random, Cnoidal and regular waves over a beach slope is brought out in Figs. 8.11, 8.12

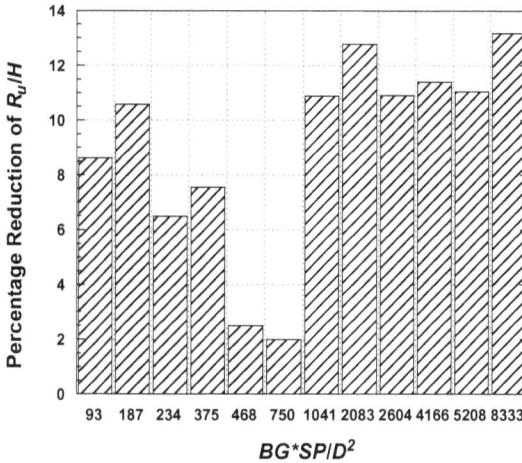

Fig. 8.13   Variation of percentage reduction of R_u/H with $BG*SP/D^2$-Regular waves.

and 8.13, respectively. In general, most of the different combinations of the characteristics of vegetation have proved to be effective in reducing the run-up. In order to have an overall picture on the aforesaid aspect, the percentage reduction in wave run-up given as the ratio of difference between run-up without vegetation and run-up with vegetation to run-up without vegetation was evaluated for all the tests. Although the run-up for Cnoidal waves was found to be a maximum, the percentage of reduction in run-up is found to be the highest for such waves compared to that of the other types of waves. Hence, this reveals that coastal plantation can serve as one of the better soft solutions at locations frequented by long waves and surges.

8.8   Summary

A well-controlled experimental programme paved the way to understand the vegetal resistance and its effects on coastal structures. An essential aspect of the study is that the hydro-elastic regime in the coastal zone pertaining to plantations along open coasts was modelled using Froude law. The effects of canopy and leaf area, wind and surface roughness of the vegetal stems were not considered.

The resistance due to vegetation in a steady-uniform flow provided in this chapter discussed Darcy–Weisbach's friction factor f and Manning's friction co-efficient n in both regular and staggered configuration of elastic cylindrical stems emerging from the water surface. The friction factors have been investigated in terms of Re, Fr, Reduced Velocity (V_r) and the VFP. The VFP is a newly proposed non-dimensional parameter to represent the relative elasticity of vegetal stems with respect to flow characteristics.

The discussions indicate that the friction factors have an inverse relationship with Fr and V_r except for the regime of V_r, where lock-in between the downstream vertices behind the cylindrical member and the natural frequency of the vegetal stems takes place. The lock-in process has been demonstrated to be effective in enhancing vegetal resistance. At this regime of V_r, the friction factors increase by about 150–200% compared to that of no-lock in.

It has been demonstrated that the VFP is effective in representing the relative elasticity of vegetal stems to that of flow characteristics. The relation between friction factors and the VFP is found to be almost linear for both regular and staggered configurations of the vegetation. Based on the earlier studies, new semi-empirical relationships are proposed for f, n and E_L (energy loss) for all the flow regimes tested in terms of Fr and VFP.

References

Armanini, A., and Righetti, M. (2002). "Flow resistance in open channel flows with sparsely distributed bushes." *Journal of Hydrology*, 269, 55–64.

Chakrabarti, S. K. (1983). *Hydrodynamics of Offshore Structures*. Computational Mechanics Publications, Southampton Boston.

Chen, C. I. (1976). "Flow resistance in board shallow grassed channels." *Journal of the Hydraulics Division, ASCE*, 102(3), 307–322.

Dudley, F. S. J., Abt, S. R., Bonham, C. D., Watson, C. C., and Fischenich, J. C. (1998). "Evaluation of flow-resistance equations for vegetated channels and floodplains." Technical Report EL-98-2, U.S. Army Engineer Research and Development Center, Vicksburg, MS.

Fathi-Moghadam, M. (2007). "Characteristics and mechanics of tall vegetation for resistance to flow." *African Journal of Biotechnology*, 6(4), 475–480.

Fischenich, C. (2000). "Resistance due to vegetation, EMRRP technical notes." ERDC TN- EMRRP-SR-07, US Army Engineer Research and Development Center, Vicksburg, MS.

Freeman, G., Rahmeyer, W., and Copeland, R. R. (2000). "Determination of resistance due to shrubs and woody vegetation." Technical Rep. No. ERDC/CHL TR-00-25, U.S. Army Corps of Engineers Engineer Research and Development Center, Vicksburg, MS.

Furukawa, A. (1997). "Stomatal frequency of Quercus myrsinaefolia grown under different irradiances." *Photosynthetica*, 34, 195–199.

Hamzah, Latief, Harada, K., and Imamura, F. (1999). "Experimental and numerical study on the effect of mangrove to reduce tsunami." *Tohoku Journal of Natural Disaster Science*, 35, 127–132.

Harada, K., and Imamura, F. (2001). "Experimental study on the resistance by mangrove under the unsteady flow." *The Proceedings of the First Asian and Pacific Coastal Engineering Conference*, 2, 975–984.

Harada, K., and Imamura, F. (2005). "Effects of coastal forest on tsunami hazard mitigation — A preliminary investigation" in Kenji Satake (ed.), *Tsunamis, Case Studies and Recent Developments*. Springer, Netherlands, 279–292.

Hiraishi, T., and Harada, K. (2003). "Greenbelt tsunami prevention in South Pacific region." Report of the Port and Airport Research Institute, 42(2).

Hu, Z., Suzuki, T., Zitman, T., Uittewaal, W., and Stive, M. (2014). "Laboratory study on wave dissipation by vegetation in combined current–wave flow." *Coastal Engineering*, 88(3), 131–142.

Isaacson, M. (1979). "Wave force on compound cylinders." Proceedings of Civil Engineering Oceans IV, 519–530.

Kouwen, N., and Fathi-Moghadam, M. (2000). "Friction factors for coniferous trees along rivers." *Journal of Hydraulic Engineering*, 126(10), 732–740.

Kouwen, N., Unny, T. E., and Hill, H. M. (1969). "Flow retardance in vegetated channels." *Journal of the Irrigation and Drainage Division, ASCE*, 95(2), 329–342.

Meijer, D. G., and Van Velzen, E. H. (1999). "Prototype-scale flume experiments on hydraulic roughness of submerged vegetation." Proceedings of XXVIII AIHR Conference, Graz (A), September.

Mendez, F. J., and Losada, I. J. (2004). "An empirical model to estimate the propagation of random breaking and nonbreaking waves over vegetation fields." *Coastal Engineering*, 51, 103–118.

Nepf, H. M. (1999). "Drag, turbulence and diffusion in flow through emergent vegetation." *Water Resources Research*, 35(2), 479–489.

Noarayanan, L. (2009). "Studies on the effect of vegetation on the forces and run-up on structures due to waves." Doctoral thesis, Indian Institute of Technology Madras, India.

Noarayanan, L., Murali, K., and Sundar, V. (2012a). "Effect of flexible vegetation on the forces with Keulegan–Carpenter number on structures due to long waves." *Journal of Marine Science and Application*, 11, 24–33.

Noarayanan, L., Murali, K., and Sundar, V. (2012b). "Role of vegetation on beach run-up due to regular and Cnoidal waves." *Journal of Coastal Research*, 28, 123–130.

Noarayanan, L., Murali, K., and Sundar, V. (2013). "Manning's 'N' for staggered flexible emergent vegetation as a possible tsunami mitigation measure." *Journal of Earthquake and Tsunami*, 7(5), 1250029 (1–18).

Petryk, S., and Bosmajian, G. (1975). "Analysis of flow through vegetation." *Journal of the Hydraulics Division ASCE*, 101(7), 871–884.

Sundar, V., Murali, K., and Noarayanan, L. (2011). "Effect of vegetation on run-up and wall pressures due to Cnoidal waves." *Journal of Hydraulic Research*, 49(4), 562–567.

Tanaka, N., Sasaki, Y., Mowjood M. I. M., Jinadasa, K. B. S. N., and Homchuen, S. (2007). "Coastal vegetation structures and their functions in tsunami protection: Experience of the recent Indian Ocean tsunami." *Landscape and Ecological Engineering*, 3(1), 33–45.

Timoshenko, S. P., and Gere, J. M. (1961). *Theory of Elastic Stability*. McGraw Hill, New York.

Werth, D. (1997). "Predicting flow resistance due to vegetation in flood-plains." Doctoral thesis, Utah State University, Logan, Utah.

Wiegel, R. L. (1964). "Tsunamis, storm surges, and harbor oscillations." Chapter 5 in *Oceanographical Engineering*, Prentice-Hall, Englewood Cliffs, N.J., 95–127.

Wu, F. C., Shen, H. W., and Chou, Y. J. (1999). "Variation of roughness coefficients for unsubmerged and submerged vegetation." *Journal of Hydraulic Engineering*, 125(9), 934–942.

Yao, Y., Tang, Z., and Jiang, C., He, W., and Liu, Z. (2018). "Boussinesq modeling of solitary wave run-up reduction by emergent vegetation on a sloping beach." *Journal of Hydro-Environment Research*, 19, 78–87.

Yao, Y., Du, R., Jiang, C., Tang, Z., and Yuan, W. (2015). "Experimental study of reduction of solitary wave run-up by emergent rigid vegetation on a beach." *Journal of Earthquake Tsunami*, 09(05), 1540003.

Chapter 9

Tsunami Generation in Laboratory

9.1 Introduction

In the last few decades, there have been considerable studies on tsunamis and their impacts on the shoreline with special focus on beaches and coastal structures through physical and numerical modelling. In particular, one of the important tasks is the description of tsunami turbulent flow, when it propagates towards the coast carrying mud, sand, stones, pieces of wood and other debris. For description of this kind of flow, one of the approaches used is based on the traditional dam breaks. However, this will represent only the breaking bore characteristics. In reality, the resulting wave may have different shapes or wave profiles from the same tsunami generation when it hits the coast. A typical sketch is provided in Shuto (1985), as shown in Fig. 9.1.

The initial wave changes in shape during its propagation towards the shore, owing to the confounding/complicated bathymetry and coastal topography. Thus, the same tsunami events lead to different manifestations at different locations. Hence, it is mandatory to study in detail the varying wave shapes. This has to be reproduced in a wave flume. The major limitations in the experimental studies on tsunamis are scaling of small wave steepness and long wave length compared to bed slope. Hence, most of the experimental testing facilities, which work using piston-type wave makers, would fail to generate longer period waves. This chapter presents the methodology that was used to generate these long-period waves as well as comparison between the real tsunami and other long wave theories like the solitary or Cnoidal waves.

9.2 Difference between Tsunami and Solitary Waves

Madsen *et al.* (2008) reported that the characteristics of solitary or Cnoidal (long waves) do not replicate the real tsunamis. However, Synolakis (1986)

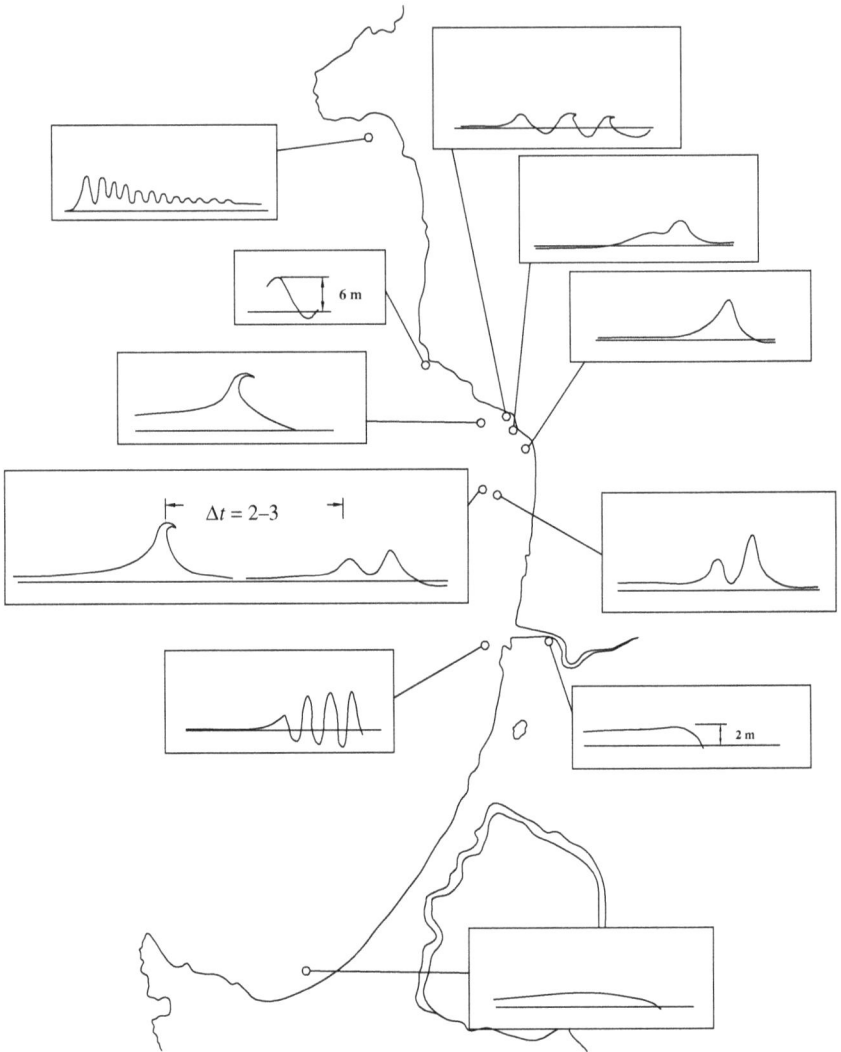

Source: Shuto (1985).

Fig. 9.1   Tsunami wave shapes at the Japancsc coast after the Japan Sea tsunami
occurred on 26 May 1983.

and Goring (1979) have stated that solitary or Cnoidal wave theories have
been adopted for modelling of tsunami propagation, shoaling and run-up.
In order to explain this in detail, let us analyse the real-field measured
tsunami and compare it with the traditional theory. Figure 9.2 shows two

measurements of the Tohoku tsunami 2011 and the Indian Ocean tsunami 2004 and their comparisons with the solitary waves. Solitary waves are surely favourable, as they have a sound theoretical background, in terms of satisfying the Korteweg de Vries equation under the assumption that nonlinearity $\varepsilon = H/d$ and dispersion ($\mu^2 = (d/L)^2$) are in balance, and over constant water depth they are stable non-periodic (transient) wave forms comparable to a tsunami.

In the solitary wave theory, temporal evolution of the water surface elevation $\eta(t)$ is described as follows:

$$\eta(t) = H \operatorname{sech}^2\left(kC(t - t_0)\right), \tag{9.1}$$

with H: wave height, $k = \sqrt{\frac{3H}{4d^3}}$: wave number, d: water depth, $C = \sqrt{g(d + H)}$: wave celerity, g: gravitational acceleration and t_0: time shift.

Thus, height is the only parameter to define its shape. However, Madsen *et al.* (2008) demonstrated that solitary waves actually do not represent a tsunami closely, not at least due to the balance of ε and μ^2 their period and length depends on the water depth, and solitary waves are unrealistically short in small water depth near the coast. This is clearly depicted in Fig. 9.2, where one can match the height for the corresponding water depth using the solitary waves; however, the period is much shorter than the real tsunami period. This is the paradigm as explained by Madsen *et al.* (2008).

Hence, one needs to model either the space–space (i.e., length and height) or space–time correctly. Shuto (1985) broadly classified tsunamis as the following:

- Non-breaking waves that act as a rapidly rising tide, observed during small and moderate tsunami events after a short distance propagation;
- Breaking bore or hydraulic jump (wall of water), observed as a result of wave breaking during large tsunami events after a short distance propagation;
- Undular bore, observed after long distance propagation (in terms of wavelength), i.e., the disintegration into series of solitons.

The initial characteristics are similar to the long sinusoidal waves, wherein the stroke required may be large and impractical to generate in the flume. For laboratory investigations, tests with different water depths could be considered. The second characteristics can be analysed by using the dam break setup. The third characteristics need a proper generation mechanism in the flume to study the propagation of such waves over a

(a)

(b)

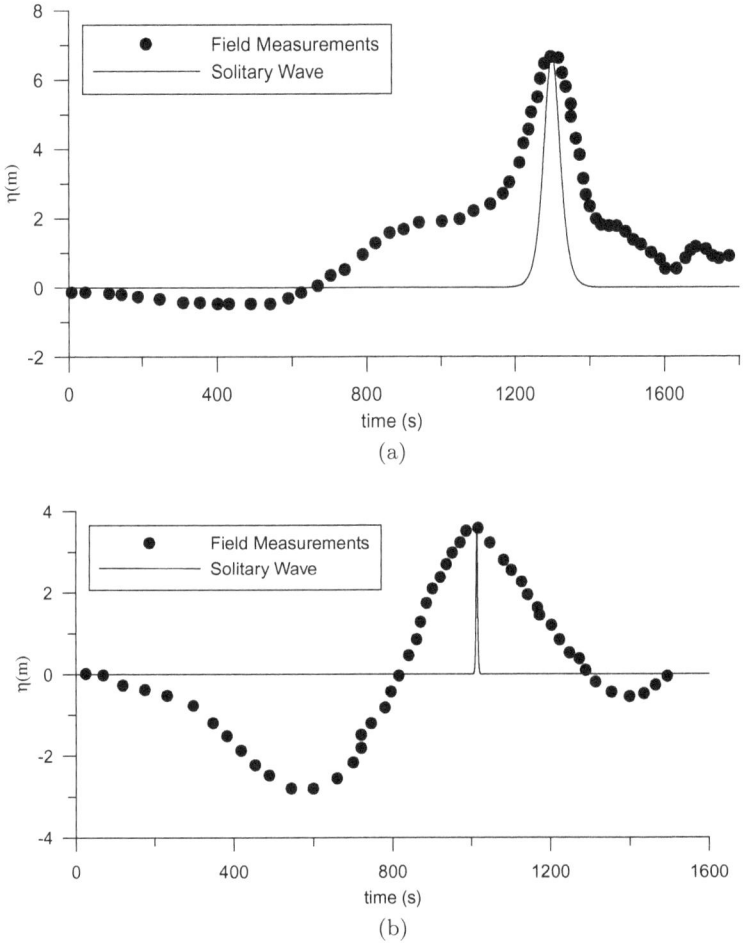

Source: Rabinovich and Thomson (2007).

Fig. 9.2   Measurement at (a) 2011 Tohoku Tsunami (Iwate South): d = 204 m; H ≈ 6.7 m; T ≈ 1000 s; L ≈ 45 km; $\varepsilon = 33 \cdot 10^{-3}$; $\mu^2 = 2 \cdot 10^{-4}$. (b) 2004 Indian Ocean Tsunami: d = 14 m; H ≈ 6.6 m; T ≈ 800 s; L ≈ 11 km; $\varepsilon = 47 \cdot 10^{-2}$; $\mu^2 = 16 \cdot 10{-}7$ (Mercator yacht).

long distance. However, most flumes have limitations in achieving such a generation. Thus, in laboratory modelling, it is difficult to generate the tsunami, and the waves studied in the laboratory should be called tsunami-like waves and this notation will be used in this chapter, and the generation methodology for the last two characteristics will be discussed.

9.3   Existing Tsunami Generator

The tsunami (bore or surge) can be reproduced in the laboratory by vertical release (Chanson *et al.*, 2002, 2003), dam-break gates (vertical or swing
type), landslide-type wave generators either manually by dropping the solid
block inside the water (Thusyanthan and Madabhushi, 2008) or mechan-
ically using a pneumatic landslide generator (Fritz, 2002), which can be
carried out only in a wetback system. The vertical release in generating a
tsunami bore involves the drop-in volume of water from the upper reservoir
to the flume bottom resulting in the formation of turbulent bores. Cur-
rently, the use of dam-break waves generated by the sudden release of the
water impounded behind the vertical gate has been justified as the more
appropriate method to reproduce the inundating tsunami inside the coast
by Chanson (2005, 2006). The dam-break setup to simulate the tsunami
bore or surge was used by several authors (Yeh *et al.*, 1989, Arnason *et al.*,
2009) because the theoretical prediction of the bores can easily be made
without any difficulty from the classical dam-break problem. The advantage
of using vertical release and the dam-break gate for the tsunami generation is the flexibility in controlling the volume of water, thereby controlling
the inundation depth and the velocity of the bore or surge. The vertical
release claims another advantage of having control over the discharge from
the overhead reservoir, by means of the passage (pipe) between the upper
reservoir and the bottom of the flume (Wüthrich *et al.*, 2018, Chanson
et al., 2002, 2003).

Recently, generation of tsunamis close to reality using a pneumatic wave
maker (Rossetto *et al.*, 2011) or a pump-driven wave maker (Goseberg
et al., 2013) have been achieved. Nevertheless, the results of the studies
have shown that several challenges still exist. In particular, from the very
beginning it was difficult to generate stable waves as the generation process
in the region close to the wave maker may suffer from high turbulence levels, breaking waves or compressed air–water phases. If the hydrodynamics
are not well defined at the point of generation, the interpretation and use
of these studies become complicated, e.g., (i) even if the water surface elevation at the point of interest (e.g., toe of a slope at the end of the flume)
looks reasonable there will always be doubts about a proper representation
of the total hydrodynamics and (ii) the data might not be suitable for the
validation of numerical models as the essential definition of well-defined
boundary conditions are difficult. Further, new generation techniques have
been installed in rather short or even bended flumes, precluding lots of
investigations, e.g., at natural shores, which are usually very mildly sloped.

Rossetto *et al.* (2011) concluded that none of the existing large-scale testing facilities, which are usually equipped with a piston-type wave maker of reasonable stroke will be able "to produce long period or trough-led waves", with periods O (100 s).

9.4 Tsunami Generation Methodology Using Piston-Type Wave Paddle

Contrary to the report of Rossetto *et al.* (2011), Schimmels *et al.* (2016) and Sriram *et al.* (2016) showed that using the wave paddle one can generate these long waves provided the sufficient stroke of the wave paddle is available. Thus, this is well suited for the validation of numerical models as it leads to well-defined boundary conditions.

The following are two approaches to generate the tsunami waves or long-period waves using the piston wave maker

(a) Traditional wave maker theory for shallow water waves
(b) Self-Correcting Method (SCM)

9.4.1 *Traditional wave maker theory*

There are two ways to obtain the stroke signal of a piston-type wave board to generate the desired water surface elevations through this method. For periodic regular or irregular wave trains the time series of the water surface elevation is transformed to frequency domain by Fast Fourier Transform (FFT), then a transfer function is applied to each wave component and the time series of the stroke signal is obtained by Inverse Fast Fourier Transform (IFFT). However, this approach works only for weakly nonlinear waves, and even if a higher-order transfer function is used (Schäffer, 1996) the generation of highly nonlinear Cnoidal waves or solitary waves will fail. Hence, for the generation of solitary waves, Goring (1979) proposed an analytical solution and Synolakis (1986) showed for small amplitude waves that instead of using Goring's theory, the time series of the wave board motion $S(t)$ can also be obtained by integrating the water particle velocity profile at the mean position of the wave board $u(0,t)$, as follows:

$$\frac{dS}{dt} = u(0,t), \qquad (9.2)$$

which for shallow water waves can be expressed in terms of the water surface elevation

$$\frac{dS}{dt} = \frac{c\eta}{d + \eta}.$$ (9.3)

Synolakis (1986) applied a fourth-order Runge–Kutta scheme for integration of Eq. (9.3), which for a solitary wave gave the same wave board motion as the method of Goring (1979). For very long small amplitude waves, i.e., $\varepsilon \gg \mu^2$ also lower-order integration methods are sufficient to obtain the stroke signal and due to its generic adoptability, this approach can be used to generate waves. Thus, in this approach one needs the target profile. The important aspect is how to model these wave forms that represent long waves or tsunamis (that have a leading depression and positive wave (Tadepalli and Synolakis, 1994)). There are three ways of obtaining it: (i) to represent it is by studying the characteristics of tsunami using long-period regular waves (Goseberg *et al.*, 2012), (ii) elongated single pulse and (iii) to represent it by sech^2(.) profile as suggested by Chan and Liu (2012). The sech^2(.) profile represents the long wave as follows:

$$\eta(t) = \sum_{i=1}^{N} H_i \operatorname{sech}^2 \omega_i (t - (t_0 + t_i)),$$ (9.4)

where H_i and ω_i corresponds to wave height and angular frequency of sech^2(.) profile that can be fitted at a location t_i to get the required profile using "i" number of such profiles, t_0 is the reference position. For example, with the aforementioned equation with chosen parameters, one can easily fit into Iwata south tsunami and Mercator time series of Fig. 9.2. The chosen parameter for Iwata south is given by Chan and Liu (2012) as $H = [-0.8\ 2.2\ 5.85]$ m, $\omega = [0.179\ 0.198\ 0.653]$ min^{-1} and $t_i = [9.67\ 16.33\ 21.63]$ min and for the Mercator time series it is taken as $H = [-3.1\ 3.8\ -1]$ m; $\omega = [0.0042\ 0.005\ 0.01]$ s^{-1} and $t_i = [600\ 1000\ 1399]$ s. These are projected in Fig. 9.3. Thus, one could easily scale down any profile to model these long-wave cases in numerical or experimental flume. It can be seen that the combination of solitons is a very reasonable approximation, when compared to the solitary and elongated waves. The elongated solitons are generated using $N = 1$ in Eq. (9.4).

   It should be noted that applying the generic approach and choosing arbitrary wave heights and periods for the solitons will result in the generation of unstable wave forms, as nonlinearity and dispersion are not in

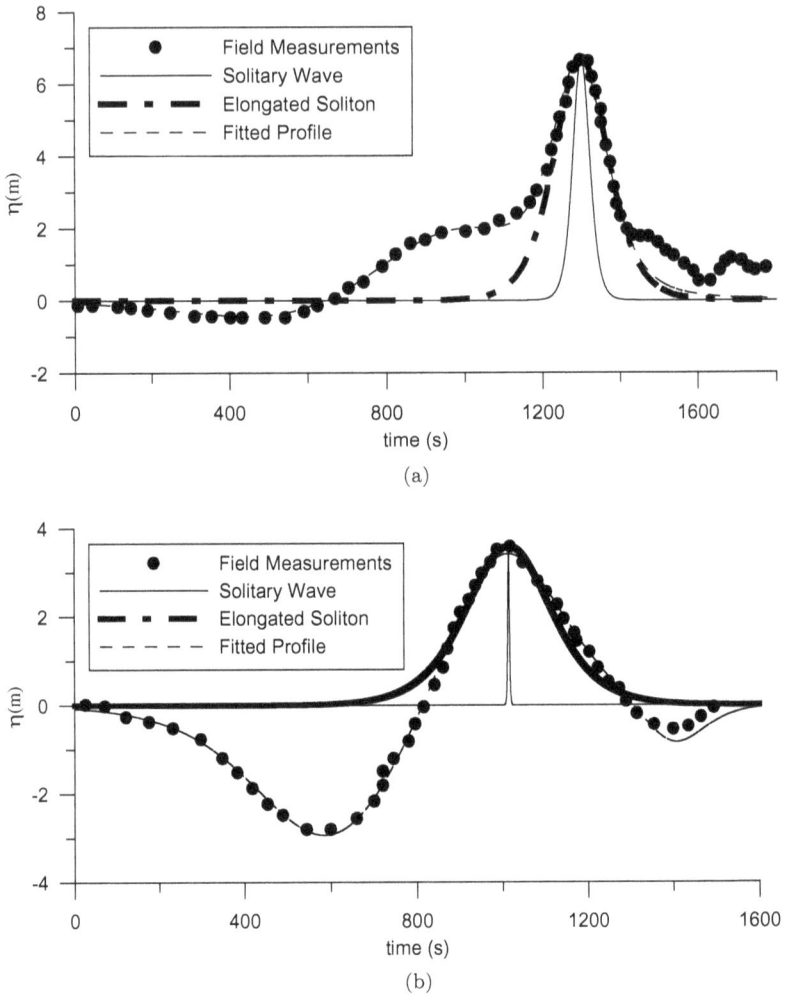

Fig. 9.3 (a) Representation of field data (Tohoku 2011 tsunami) using sech^2 waves (Chan and Liu, 2012). (b) Representation of field data (Indian Ocean 2004 tsunami) using sech^2 waves.

balance. However, as clearly shown by Madsen *et al.* (2008) the transformation of very long waves and the generation of shorter waves riding over the long wave is a natural process and could also be observed for tsunamis, which are tend to become asymmetric and are likely to transform to an undular bore while approaching the shallow waters of the continental shelf. Thus, one can generate the stroke through the target profile using numerical integration.

9.4.2 Self-correcting method

The best way to improve the quality of the control signal is by employing an SCM. It accounts for the inherent nonlinear interactions in the resultant wave train and thereby improving the efficiency of wave focussing. The pioneering work on this method has been done in the 1980s by Daemrich et al. (1980) or higher-order regular waves (Daemrich and Götschenberg, 1988) by establishing amplitude and phase-transfer functions. However, this method may not always work due to the generated spurious, free sub- or super-harmonic components, which have an influence in reaching the target wave profile using linear paddle displacement or due to the unrealistic assumption of the target wave profile. Later, a similar approach for focussed wave generation applying phase corrections alone was adopted by Chaplin (1996) that yielded good results. Recently, Schmittner et al. (2009) with corrections for amplitude and phase, excellent results were exhibited for improving the deterministic wave trains. It is well known that the steep focussed wave based on the linear theory cannot practically be reproduced in the laboratory wave flume, and it was reported that wave–wave interactions for producing these focussed waves are interactions of third order (Baldock et al., 1996). Hence, the target signal should be based on higher-order theory rather than linear theory. Figure 9.4 illustrates the algorithm of the SCM using a numerical model.

A brief overview has been given in the following paragraphs; more details and application to variable water depth can be referred in Fernández et al. (2013). The performance of the method using different correction schemes are reported by Fernández et al. (2014).

An initial control signal using the principle of wave superposition as stated earlier by identifying the target signal is generated. The first estimation of the stroke has to be made from the field measurements. The simulation is then carried out using the numerical techniques or experiments and the free surface is recorded at the location, where one needs the target profile. Now, the signal is compared in the frequency domain with the target signal. On comparing the amplitudes and phases, the new components of the control signal are extracted as follows; the phase shift between the theoretical and the measured wave profile is computed by the

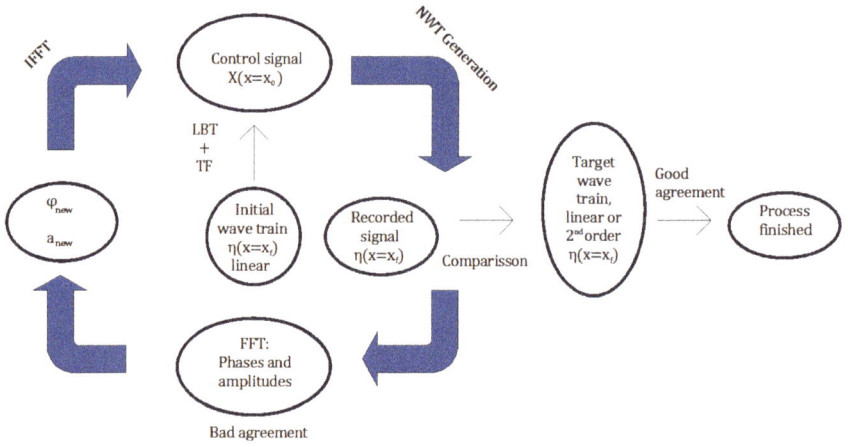

Note: FFT: Fast Fourier Transform; IFFT: Inverse Fast Fourier Transform; LBT: Linear Back Transformation; TF: Transfer Function.

Fig. 9.4   Illustration of the self-correcting algorithm.

following:

$$\varphi_s = \varphi_{\text{target}} - \varphi_{\text{recorded}}, \tag{9.5}$$

with φ_s being the phase shifts between φ_{target}, the phase angles of the target focussed wave at the target location, and $\varphi_{\text{recorded}}$, the phase angles of the recorded focussed wave at the focal point. The new phase angles for the control signal are calculated applying the calculated phase shift to the phase angles of the control signal simulated previously.

$$\varphi_{\text{new}} = \varphi_{\text{old}} \pm \varphi_s, \tag{9.6}$$

where φ_{new}, are the new phase angles for the new control signal and φ_{old} are the phase angles of the previous control signal; so in the expression presented Eq. (9.6), the phase shift should be subtracted to the old phase angles if the phase shift is negative and *vice versa*.

Apart from the phase corrections, one can also correct the amplitudes of the previous control signal taking into account the ratio between the amplitudes of the recorded wave and the theoretical one (Daemrich *et al.*, 1980):

$$a_{\text{new}} = a_{\text{old}} \times \left(\frac{a_{\text{target}}}{a_{\text{recorded}}} \right), \tag{9.7}$$

where a_{new} are the new amplitudes of the wave component in the new control signal, a_{target} are the amplitudes of the theoretical focussed wave at the focal point and a_{recorded} are the amplitudes of the recorded wave component at the focussing location. Note that this is the amplitude spectrum and not the spectral density, and an inverse Fourier transform is applied after the corrections. The spectrum of the new wave profile to be generated is corrected step by step to fit the spectrum of the theoretical wave. Thus, the SCM is applicable with a minimum knowledge of the wave generation system (electronics and mechanical parts) and can be used for flap or piston or combination of flap and piston type and even for the pneumatic wave generation and tide generation to reproduce the tsunami time series in the flume (Rossetto et al., 2011; Goseberg et al., 2013).

9.5   Limitation of the Method

Basically, the generation of solitary or tsunami-like waves is just a problem of the displacement of a certain water volume, be it with water pumps, air pressure or a wave board. The volume under a solitary wave is given as follows:

$$V = \frac{1}{\pi}\sqrt{gd}HT. \tag{9.8}$$

The maximum wave board stroke follows easily from

$$S_{\text{max}} = \frac{V}{d} = \frac{1}{\pi}\sqrt{\frac{g}{d}}HT. \tag{9.9}$$

These simple algebraic expressions allow for an easy assessment of the sufficient volume of water to be displaced and the necessary stroke to generate a certain solitary wave. Figure 9.5 shows two contour plots for the water volume per metre width (Fig. 9.5(a)) and the corresponding stroke (Fig. 9.5(b)) depending on the water depth d and the product of wave height and period HT. The bold line in Fig. 9.5(b) marks a typical stroke that is available in most of the generation testing facility around the world, i.e., a stroke of 1.6 m. This was marked in order to point out the possibilities for tsunami-like generation in wave flume with the limitations of the available wave maker. The arrows indicate the maximum achievable combination of wave height and period, HT = 1.7, for the experiments that can be carried out in a water depth of 60 cm. Similar comparison for the large-scale testing facility, for example, Großer Wellenkanal, GWK, of Forschungszentrum Küste (FZK), Germany is reported by Schimmels et al. (2016).

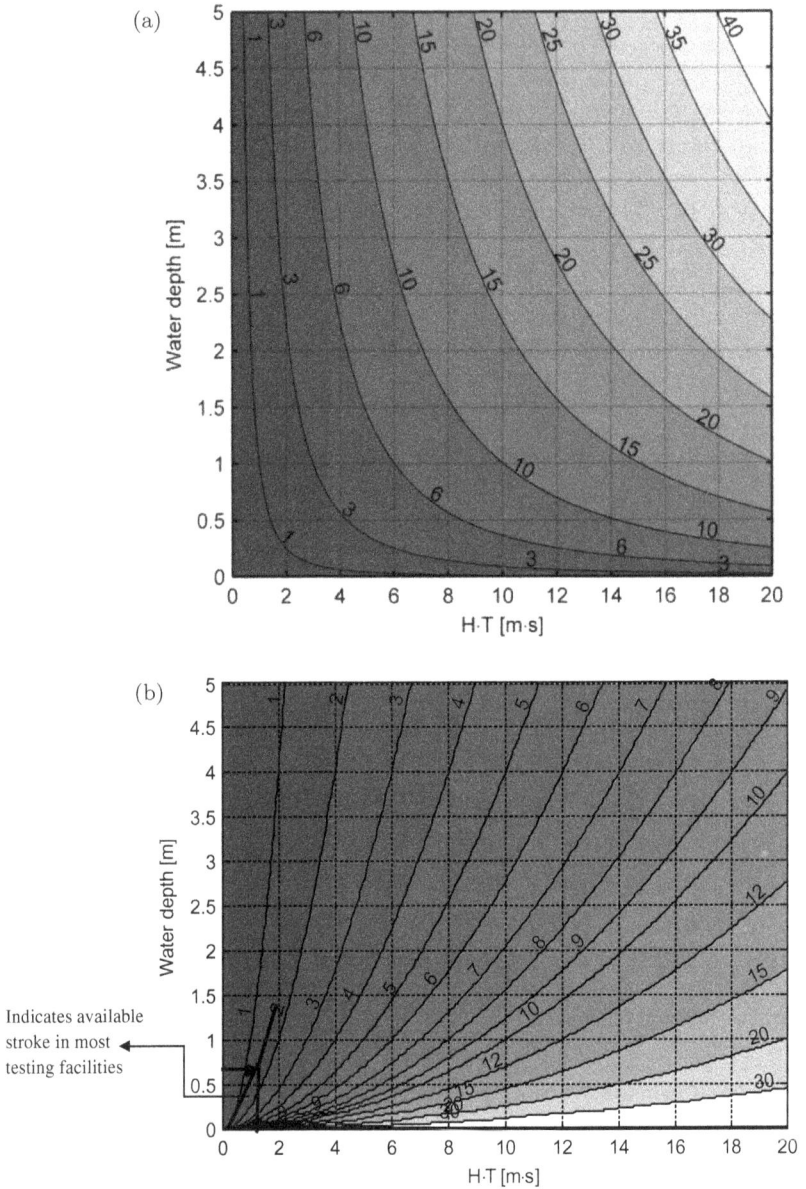

Fig. 9.5   Necessary (a) water volume [m³/m] to be displaced and (b) wave board stroke [m] for solitary wave generation.

9.6 Illustration of the Aforesaid Two Methods Using Experiments and Numerical Model

The large-scale experimental study was carried out at the Large Wave Flume (GWK). A detailed description of the experimental setup and the generation methodology is discussed by Schimmels *et al.* (2016) and will not be repeated here. However, for the completeness of this chapter, the tested cases are extracted and shown in Table 9.1. For the generation of the waves, Schimmels *et al.* (2016) used a generic approach by combining several solitons (sech^2-waves) as suggested by Chan and Liu (2012).

Additionally, another method based on an SCM as described by Fernández *et al.* (2014) was adopted, through which downscaled sea-level oscillations, measured in Pago Pago harbour during Samoa 2009 tsunami, is reproduced.

In this section, the results and discussion of the previous test cases are reported along with the numerical modelling based on FEM (herein, referred as, FNPT–FEM developed by Sriram *et al.* (2006)) are reported. The aforementioned test cases are classified into four different categories, *viz.*, regular wave, positive impulse, combination of three sech^2 profile and the reproduction of tsunami time series using the SCM. However, only positive impulse or elongated soliton, N waves and reproduction of tsunami time series are reported in this chapter as an example.

9.7 Elongated Solitary-Like Waves

In this section, the solitary wave and long positive pulses (elongated solitary-like waves) are reported. The experimental investigations on solitary waves have been carried out by several researchers in the past. A typical solitary wave simulation compared with the FNPT–FEM is shown in Fig. 9.6. The figure shows a good reproducibility between the experimental and FNPT–FEM model at a distance 60 m from the wave paddle. Nevertheless, as discussed earlier, the tsunami period in nature would be much longer, hence the simulation of an arbitrary profile of the form in Eq. (9.4) with period of about 20 s is carried out.

The experimental and FNPT–FEM simulations for the test cases 2/2 and 2/3 of Table 9.1 are shown in Fig. 9.7, which correspond to waves with a period of 20 s and heights of 0.12 m and 0.06 m. The corresponding strokes for the abovementioned cases are also projected in the figure. Further, one could clearly see that the profiles are stable in the present methodology for long wave generation up to 60 m from the wave paddle. However, at

Table 9.1.   Tested wave characteristics in GWK facility.

Test no	H1 (m)	H2 (m)	H3 (m)	T1 (s)	T2 (s)	T3 (s)	t1 (s)	t2 (s)	t3 (s)	Type
1/1	0.06	—	—	100	—	—	—	—	—	Regular wave
1/2	0.08	—	—	60	—	—	—	—	—	Regular wave
2/1	0.04	—	—	30	—	—	—	—	—	1 sech^2 profile (elongated soliton)
2/2	0.06	—	—	30	—	—	—	—	—	1 sech^2 profile (elongated soliton)
2/3	0.12	—	—	30	—	—	—	—	—	1 sech^2 profile (elongated soliton)
2/4	0.11	—	—	30	—	—	—	—	—	1 sech^2 Profile (elongated soliton)
2/5	0.1	—	—	30	—	—	—	—	—	1 sech^2 Profile (elongated soliton)
2/6	0.05	—	—	30	—	—	—	—	—	1 sech^2 Profile (elongated soliton)
3/1	−0.02	0.04	—	100	100	—	90	135	—	2 sech^2 profile (asymmetry)
3/2	−0.04	0.04	—	100	100	—	90	135	—	2 sech^2 profile (symmetry)
3/3	−0.05	0.13	—	100	100	—	20	30	—	2 sech^2 profile (N-wave)
3/4	−0.05	0.07	—	100	100	—	20	30	—	2 sech^2 profile (N-wave)
4/1	−0.05	0.09	0.13	21	21	18	20	30	35	3 sech^2 profile (N-wave)
4/2	−0.05	0.05	0.07	21	21	18	20	30	35	3 sech^2 profile (N-wave)
5/1	−0.02	0.028	0.04	150	137	41	83.4	111.6	133.8	Iwate profile
5/2	−0.03	0.04	0.01	126	126	63	60	101	140	Mercator profile
6/1	0.12			120						Pago Pago, SCM (Fernández et al., 2014)

Source: Sriram et al. (2016).

a distance of 225 m from the wave paddle, the results from experiments and FNPT–FEM simulations for the steep waves demonstrate splitting. The second wave that is present in the experiments corresponds to the reflection from the slope. The split up of single wave to multiple waves are more pronounced for the wave height of 0.12 m compared to a wave height of 0.06 m.

The wave breaking distance X, where the wave commences to split, can be estimated from the nonlinear shallow water theory as follows

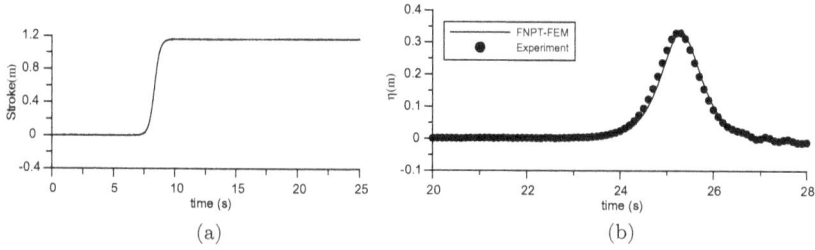

Fig. 9.6 Experimental and FNPT–FEM comparison of the solitary wave simulation. (a) Paddle stroke. (b) Simulated results at $x = 60$ m.

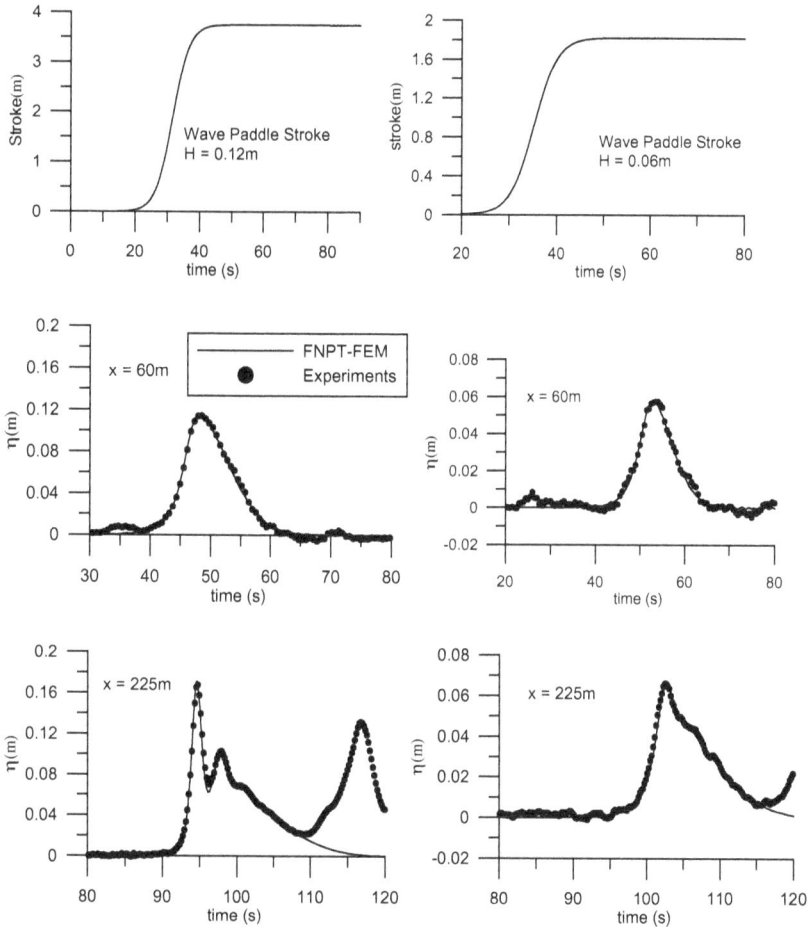

Fig. 9.7 Comparison between experiments and FNPT–FEM wave time history for positive wave pulse of 20 s period.

(Didenkulova *et al.*, 2006; Zahibo *et al.*, 2008):

$$X = \frac{\sqrt{gd}}{\max(-dV/dt)} \quad V = 3\sqrt{g(d+\eta)} - 2\sqrt{gd}. \tag{9.10}$$

The estimates of wave-breaking distance computed using Eq. (8.10), for the time series record at 50 m projected in Table 9.2, exhibit a reasonable agreement, in order to investigate further, as the FNPT–FEM model and experimental measurements are identical, and due to lack of data at different locations in experiments. The various locations of FNPT–FEM (with the spacing of 20 m) are plotted in Fig. 9.8(a) and 9.8(b) for typical two cases. These results show that the splitting of waves occur at about 150 m for 0.12 m wave height, 250 m for 0.06 m. This is in a good agreement with the estimates based on Eq. (9.10) (see Table 9.2). The wave splitting or instability in the wave propagation from experiments may be attributed to the improper paddle motion, i.e., generation of long-period wave leads to spurious harmonics generated by paddle motion; hence, one needs to go for second-order paddle motion as pointed by Schäffer (1996). However, for the tested long-period waves, the splitting is in agreement with the theoretical estimates and hence the generation algorithm based on RK fourth order for obtaining the paddle displacement is very stable.

The split up of initial long wave into solitons is well described by KdV equation. For the initial wave described by Eq. (8.4) the number of formed solitons and their wave heights, H_n can be evaluated from the following

Table 9.2. FNPT–FEM and theoretical splitting locations.

Wave height (m)	Appr. FNPT splitting Location (m) for time series recorded at 50 m	SWT splitting location (m) from wave paddle using time series recorded at 50 m	Type of wave
0.12	146	147	Positive pulse
0.06	250	251	Positive pulse
0.04	—	341	Positive pulse
0.06	—	370	Asymmetric N wave
0.08	—	346	Symmetric N wave
0.12	100	99	Two-Sech^2 (.) – A1
0.07	160	159	Two-Sech^2 (.) – A2
0.07	195	193	Three-Sech^2 (.) – A2
0.12	115	115	Three-Sech^2 (.) – A1

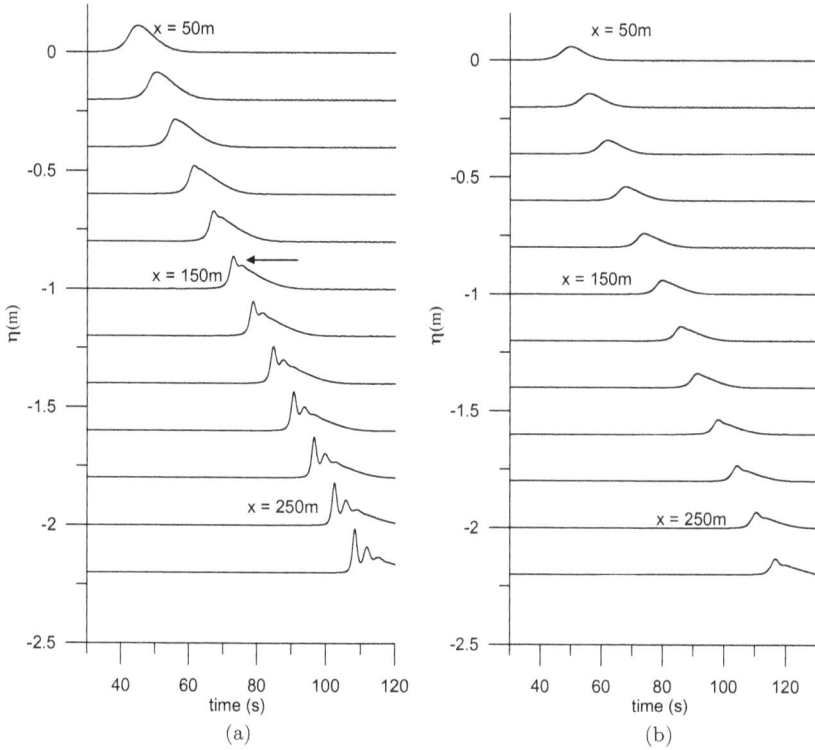

Fig. 9.8 (a) Wave propagation at different locations depicting the splitting of the waves for $H = 0.12$ m, $T_{per} = 20$ s impulse waves (the arrow represents the splitting locations which is in agreement with the theoretical splitting). (b) Wave Propagation at different locations depicting the splitting of the waves for $H = 0.06$ m, $T_{per} = 20$ s impulse waves.

equation (Pelinovsky, 1996):

$$H_n = \frac{H}{4Ur} \left[\sqrt{1 + 8Ur} - (2n - 1) \right]^2 ,$$

$$n = 1, 2, \ldots N, \ N = \left\lfloor \frac{\sqrt{1 + 8Ur} + 1}{2} \right\rfloor , \qquad (9.11)$$

where Ur is the Ursell number. It is evident from Eq. (9.11) that amplitudes of formed solitons cannot exceed twice amplitude of the initial wave. For instance, for a wave with $H = 0.12$ m, the largest soliton has an amplitude of 0.19 m. This is in agreement with the FNPT–FEM model, wherein the first wave height after 270 m is about 0.185 m.

9.8   Waves of Variable Polarity with a Leading Trough (N Waves)

It has been noticed and demonstrated by Tadepalli and Synolakis (1994) that the wave of variable polarity with a leading trough (so-called N wave) leads to a larger wave amplification and run-up over a beach. Later, this effect has been interpreted by (Didenkulova *et al.*, 2007) as an influence of face wave front steepness. Another reason as to why N waves became popular in studies related to tsunamis is their frequent occurrences (Tadepalli and Synolakis, 1996; Madsen *et al.*, 2008), which have been recently proven for distant tsunamis as an effect of elasticity of the solid earth, compressibility of seawater and the gravity potential change associated with the mass motion during the passage of the tsunami (Watada *et al.*, 2013).

In the present study, generation and propagation of two types of N waves: symmetric and asymmetric are considered. The symmetric N wave consists of two symmetric $\text{sech}^2(.)$ profiles of opposite polarity. The asymmetric N wave has a crest twice larger compared to its trough. The stroke of the paddle signal for these two N waves and the experimental measurements at $60\,\text{m}$ from the wave paddle are shown in Fig. 9.9. The results

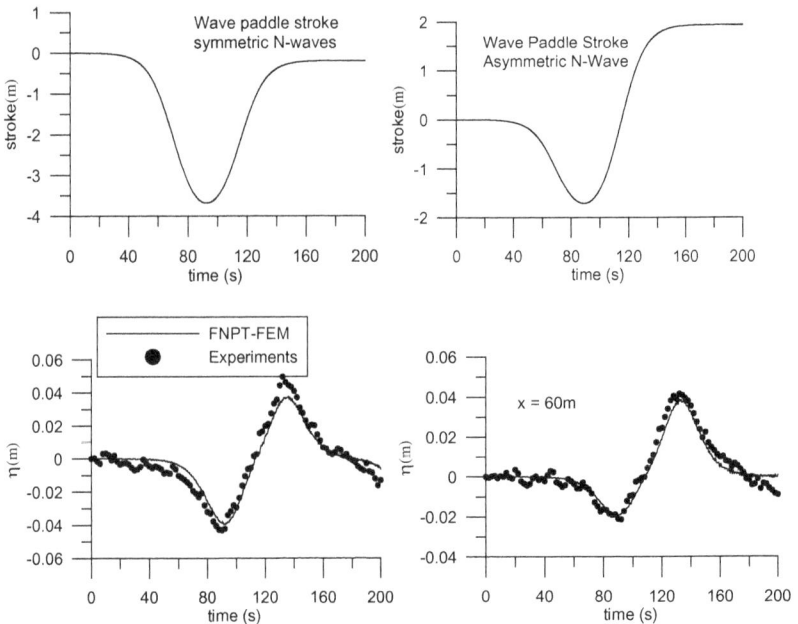

Fig. 9.9   Symmetric and asymmetric N-wave generation.

show a good agreement between the FNPT–FEM and experimental measurements. The duration or the period of the wave is about 120 s. Due to large wavelength, the wave gauges located close to the end of the flume, register also wave reflection from the beach which are not reproduced. Further, both in the FNPT–FEM and experimental measurements, no splitting is noticed and the theoretical splitting occurs only at 350 m (see Table 9.1).

The characteristics of N wave are studied by decreasing its period and increasing its heights to study its evolution during propagation. Initially, two sech^2 (.) profiles are being used and two different wave heights were tested for the same wave period. This corresponds to test case no: 3/3 and 3/4. The wave paddle stroke and wave record at 60 m and 225 m are shown in Fig. 9.10. As anticipated, a wave of larger amplitude splits sooner and into a larger number of solitary waves compared to the wave of smaller amplitude which splits into only two solitary waves. The wave evolution at different locations over the time based on the FNPT–FEM simulation are shown in Fig. 9.11(a) and 9.11(b). The approximate splitting location shown by the arrows is in agreement with the estimates performed within nonlinear shallow water theory (Table 9.2). For every water displacement in shallow water, the wave will transform into the set of solitons of different length and amplitude. Hence, the decrease in wave trough is associated with the same process. After a very long propagation, the trough should vanish and the wave profile will be represented by a set of solitons (solitary waves of positive polarity).

9.9 "Mercator" Tsunami Time Series (Three Sech^2 Profile)

The waves generated in the experimental facilities so far had been carried out only for a short period (only 30 s) and there is scope for much longer earthquake tsunamis to be reproduced in the large wave flume by a piston-type wave maker. The recording of the Indian Ocean tsunami in 2004 was measured and logged in a yacht Mercator, which was approximated by three sech^2 waves and scaled down to 1:100. This resulted in wave height of 6.4 cm and approximate total period of 120 s, a water depth of 1 m was maintained although the actual derived water depth was 14 cm. Nevertheless, the deviation in water depth is acceptable to demonstrate the capabilities of the piston-type wave maker. The wave board motion and water surface elevations at three different spatial locations in the flume is given in Fig. 9.12. The measurements obtained from GWK, NWT and also the scaled-down field measurements have been plotted in the same graph for comparison.

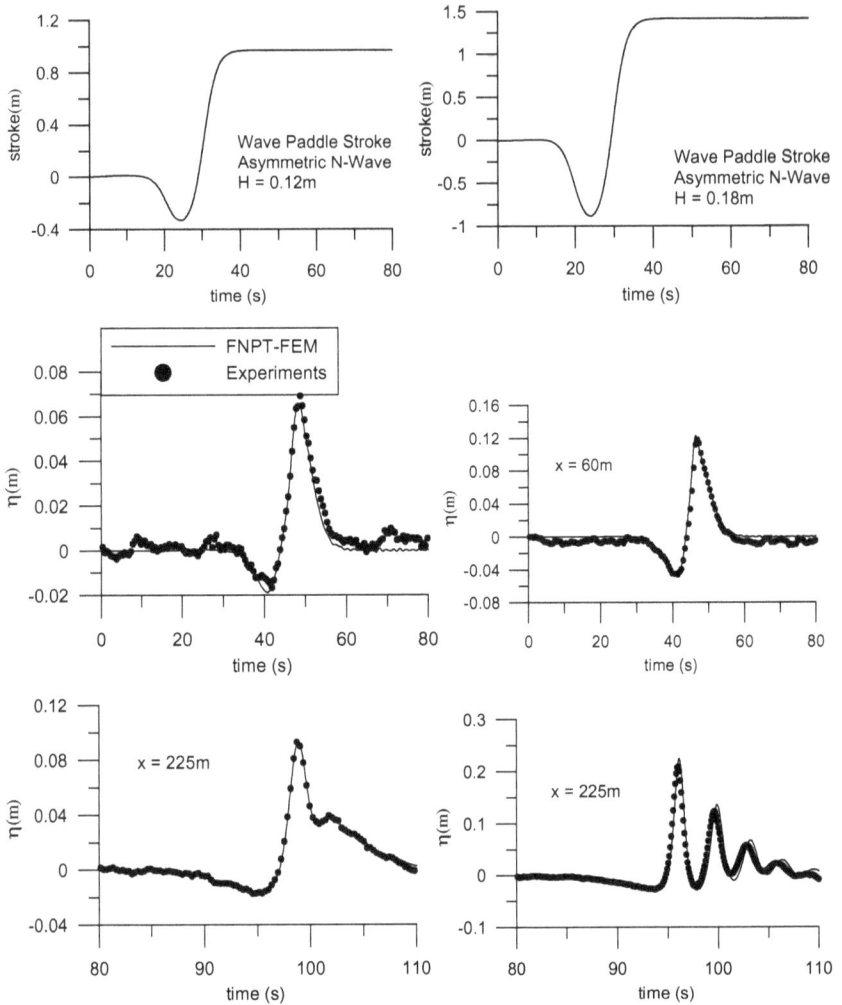

Fig. 9.10   Propagation of the two Sech^2(.) profile for long distance.

The wave board moves with a maximum stroke length of 3.6 m, which is the limit of GWK wave maker. To adapt this stroke length the 1:100 scale is adopted for this particular real tsunami time series. In this context, had the water depth been reduced, the maximum achievable wave height would be smaller. The agreement of results between GWK and NWT data is appreciably good, also the data fit with scaled-down field measurements are convincing. The agreement of results at $x = 0$ and $x = 50$ is substantial

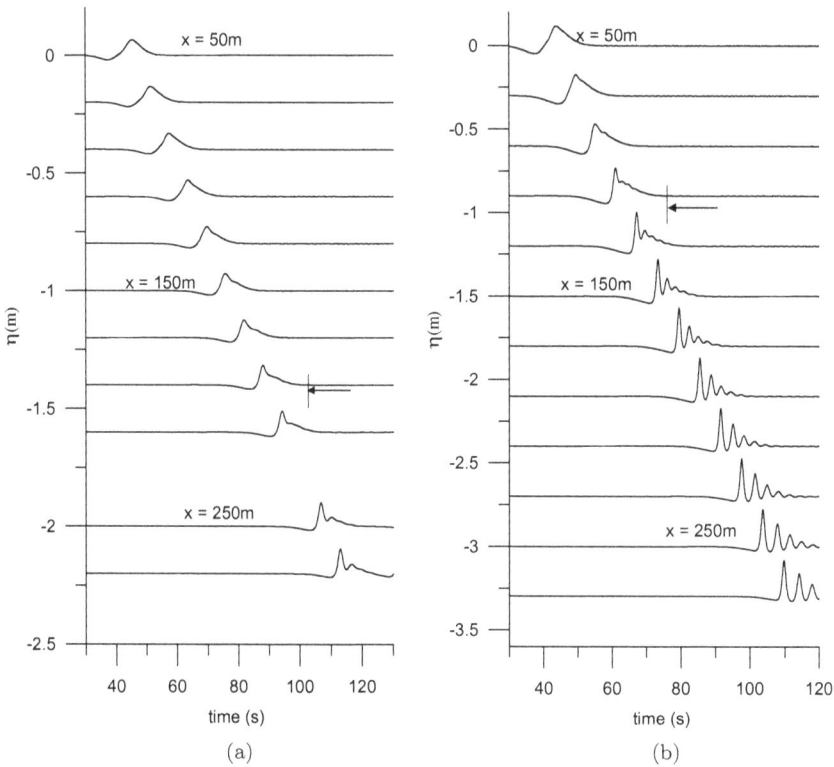

Fig. 9.11 (a) FNPT–FEM simulation of the wave propagation at different locations, showing the splitting of the wave forms. The arrow represents the numerical splitting locations. Asymmetric N Waves $H = 0.12$ m. (b) FNPT–FEM simulation of the wave propagation at different locations, showing the splitting of the wave forms. The arrow represents the numerical splitting locations. Asymmetric N Waves $H = 0.18$ m.

and when $x = 225$ certain clear deviations are observed, which are due to the reflections from the flume (1:6 slope at the flume end) walls that were absent in the field. Thus, for conducting tsunami experiments, sufficiently long flumes, i.e., at least in the order of one wave length is required.

9.10 "Pago Pago" Tsunami Time Series (SCM)

Based upon the results already obtained, it could be assumed that the tsunami generation requires a large-scale representation in a long flume. Generation of tsunami-like waves is possible to arrive as a combination of solitons (sech^2 waves) with the aid of a piston-type wave maker. The

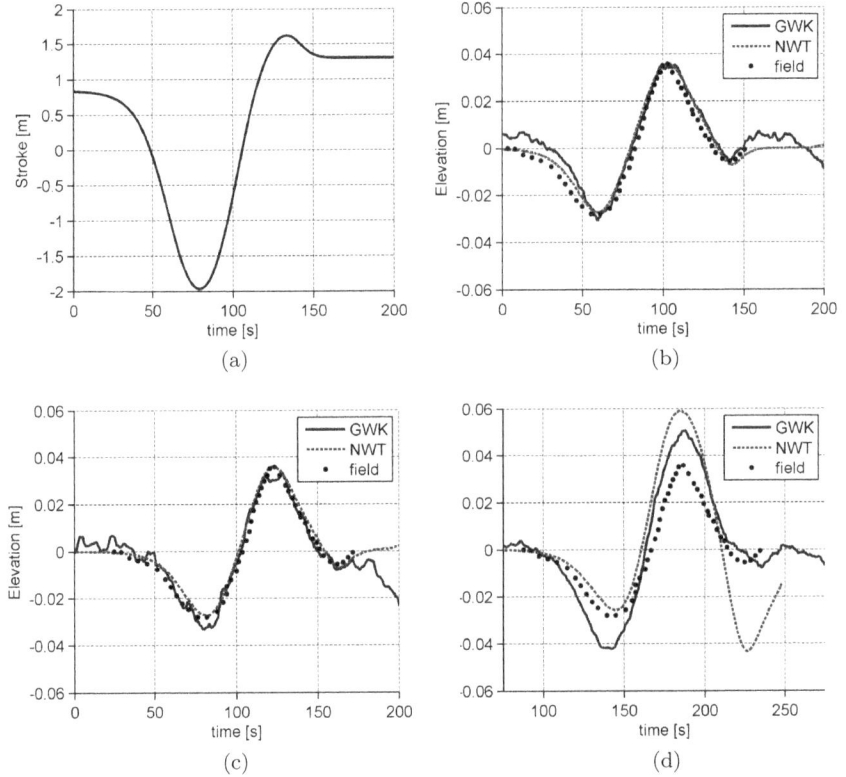

Fig. 9.12   Indian Ocean tsunami 2004 recorded by yacht Mercator on a 1:100 scale (except for water depth) approximated by a 3 sech^2 profile. (a) Paddle stroke. (b) Water surface elevation at $x = 0$. (c) Water surface elevation at $x = 50$ m. (d) Water surface elevation at $x = 225$ m.

methodology discussed herein generates water surface elevations directly at the wave board and the wave forms undergo certain transformation along its propagation length. For researchers interested in investigating a particular wave at a given location in the flume or the slope, this method is not satisfactory. Therefore, for tsunami generation the application of SCM was employed to generate the measurements from Samoa tsunami of 2009, recorded by a tide gauge at Pago Pago harbour (Zhou *et al.*, 2012; Didenkulova, 2013). The water surface elevation was scaled down to 1:37, again disregarding the water depth, and the target location was chosen to be at $x = 225$ m. The iteration has been done with the numerical wave tank and after two correction steps the solution converged. The obtained wave

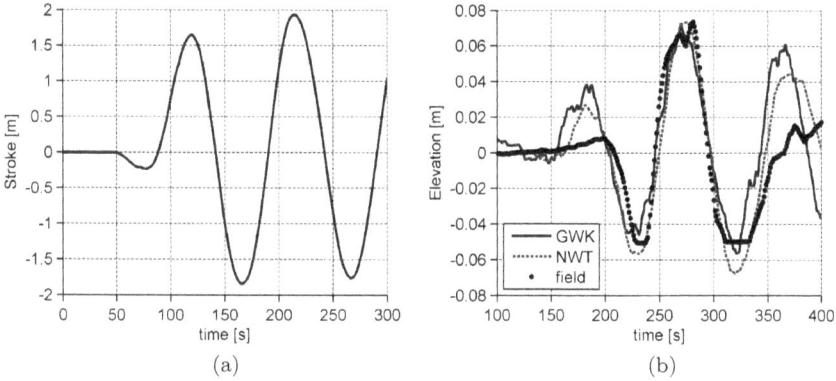

Fig. 9.13   Samoa tsunami 2009 recorded in Pago Pago harbour on a 1:37 scale (except for water depth) generated with the SCM. (a) Stroke and (b) water surface elevation at the target location, $x = 225$ m.

board motion has then been applied in GWK and is shown in Fig. 9.13 together with the results at the target location.

The stroke (Fig. 9.13(a)) again is more or less at the limit of the wave maker, which explains the rather odd scale of 1:37, which has been chosen by intention in order to present the maximum achievable scale in GWK for this particular wave. Actually, the wave board motion is reminiscent of a periodic wave rather than a transient wave event like a tsunami and in fact this is also reflected in the water surface elevation (Fig. 9.13(b)), which for the numerical simulation and the GWK data shows significantly overemphasized wave crests before and after the main wave compared to the field measurements. Furthermore, the agreement between the measurements in GWK and the NWT results is considerably worse compared to the cases before, although the same wave board motion has been used. It could be argued again that this is due to the different boundary conditions at the end of the flume (slope versus sponge layer), but this has not been investigated further nor have the deviations from the target signal. As the agreement with the main wave is reasonable it can be concluded that the SCM in principle might be applied also for tsunami wave generation in the laboratory, but there are obviously still some open issues left for optimization.

9.11   Summary

In this chapter, the tsunami generation in the laboratory has been discussed and reported. The chapter showed with a real recorded tsunami that the solitary wave assumption is not correct and any other form have to be assumed as tsunami-like waves instead of tsunami, while testing in the lab scale. Different methods of generation attempted by the researchers are discussed and the tsunami-like generation using a piston-type wave paddle has been reported. Two different methods of generations using a piston-type wave paddle have been discussed, one using the traditional method by knowing the target profile and the second method based on the iterative scheme, namely SCM. The physics of the propagation of the undular bore over long distance has been validated by comparing with analytical and numerical models, showing that the generation method is a promising approach for the tsunami waves. Nevertheless, the limitation of this generation mechanism is that the velocity of the generated wave will be less due to the wave paddle limitations. So, modelling bore-type tsunami wave dam-break experiments will be an option.

References

Arnason, H., Petroff, C., and Yeh, H. (2009). "Tsunami bore impingement onto a vertical column." *Journal of Disaster Research*, 4(6), 391–403.

Chan, I. C., and Liu, P. L. F. (2012). "On the runup of long waves on a plane beach." *Journal Geophysical Research*, 117, C08006.

Chang, Y. H., Hwang, K. Sh., and Hwung, H. H. (2009). "Large-scale laboratory measurements of solitary wave inundation on a 1:20 slope." *Coastal Engineering*, 56(10), 1022–1034.

Chanson, H. (2005). "Le Tsunami du 26 décembre 2004: Un phénoméne hydraulique d'ampleur internationale." *Premiers Constats. La Houille Blanche*, (2), 25–32.

Chanson, H. (2006). "Tsunami surges on dry coastal plains: Application of dam break wave equations." *Coastal Engineering Journal*, 48(04), 355–370.

Chanson, H., Aoki, S. I., and Maruyama, M. (2002a). "An experimental study of tsunami runup on dry and wet horizontal coastlines." *Science of Tsunami Hazards*, 20(5), 278–293.

Chanson, H., Aoki, S. I., and Maruyama, M. (2002b). "Unsteady air bubble entrainment and detrainment at a plunging breaker: Dominant time scales and similarity of water level variations." *Coastal Engineering*, 46(2), 139–157.

Chaplin J.R. (1996). "On frequency-focusing unidirectional waves." *International Journal of Offshore and Polar Engineering*, 6(2), 131–137.
remove not cited

Didenkulova, I., Pelinovsky, E., Soomere, T., and Zahibo, N. (2007). "Runup of nonlinear asymmetric waves on a plane beach" in A. Kundu (ed.), *Tsunami & Nonlinear Waves*, Springer, Berlin, 175–190.

Didenkulova, I., Zahibo, N., Kurkin, A., and Pelinovsky, E. (2006). "Steepness and spectrum of a nonlinearly deformed wave on shallow waters." *Izvestiya, Atmospheric and Oceanic Physics*, 42(6), 773–776.

Didenkulova, I. (2013). "Tsunami runup in narrow bays: The case of Samoa 2009 tsunami." *Natural Hazards*, 65(3), 1629–1636.

Fernández, H, Schimmels, S., and Sriram V. (2013). "Focused wave optimization by means of a self-correcting method." Proceedings of the 22nd International Offshore and Polar Engineering Conference, Alaska, USA.

Fernández, H., Sriram, V., Schimmels, S., and Oumeraci, H. (2014). "Extreme wave generation using self-correcting method — Revisited." *Coastal Engineering*, 93, 15–31.

Fritz, H. M. (2002). "Initial phase of landslide generated impulse waves." Doctoral dissertation, ETH Zurich.

Goring, D. G. (1979). "Tsunamis — The propagation of long waves onto a shelf." Doctoral Thesis, California Institute of Technology, Pasadena.

Goseberg, N., Wurpts, A., and Schlurmann, T. (2013). "Laboratory-scale generation of tsunami and long waves." *Coastal Engineering*, 79, 57–74.

Madsen, P. A., Fuhrman, D. R., and Schäffer, H. A. (2008). "On the solitary wave paradigm for tsunamis." *Journal of Geophysical Research*, 113, C12012.

Pelinovsky, E. (1996). "Hydrodynamics of tsunami waves." IPF RAN, Gorky [in Russian].

Rabinovich, A. B., and Thomson, R. E. (2007). "The 26 December 2004 Sumatra tsunami: Analysis of tide gauge data from the World Ocean Part 1. Indian Ocean and South Africa." *Pure and Applied Geophysics*, 164, 261–308.

Rossetto, T., Allsop, W., Charvet, I., and Robinson, D. I. (2011). "Physical modelling of tsunami using a new pneumatic wave generator." *Coastal Engineering*, 58(6), 517–527.

Schäffer, H. A. (1996). "Second-order wavemaker theory for irregular waves." *Ocean Engineering*, 23, 47–88.

Schimmels, S., Sriram, V., and Didenkulova, I. (2016). "Tsunami generation in a large scale experimental facility." *Coastal Engineering*, 110, 32–41.

Schmittner, C., Kosleck, S., and Hennig, J. (2009). "A phase-amplitude iteration scheme for the optimization of deterministic wave sequences." 28th International Conference on Ocean, Offshore and Arctic Engineering (OMAE 2009), ASME, 653–660.

Shuto, N. (1985). "The Nihonkai–Chuubu earthquake tsunami on the north Akita coast." *Coastal Engineering Japan, JSCE*, 28, 255–264.

Sriram, V., Sannasiraj, S. A., and Sundar, V. (2006). "Numerical simulation of 2D nonlinear waves using finite element with cubic spline approximation." *Journal of Fluids and Structures*, 22(5), 663–681.

Sriram, V., Didenkulova, I., Sergeeva, A., and Schimmels, S. (2016). "Tsunami evolution and run-up in a large scale experimental facility." *Coastal Engineering*, 111, 1–12.

Synolakis, C. E. (1986). "The runup of long waves." Doctoral Thesis, California Institute of Technology, Pasadena.

Tadepalli, S., and Synolakis, C. (1996). "Model for the leading waves of tsunamis." *Physical Review Letters*, 77(10), 2141–2144.

Tadepalli, S., and Synolakis, C. E. (1994). "The runup of N-waves." *Proceedings of the Royal Society of London A*, 445, 99–112.

Thusyanthan, N. I., and Gopal Madabhushi, S. P. (2008). "Tsunami wave loading on coastal houses: A model approach." *Proceedings of the Institution of Civil Engineers-Civil Engineering, Thomas Telford Ltd.*, 161(2), 77–86.

Watada, S., Kusumoto, S., and Satake, K. (2013). "Cause of travel-time difference between observed and synthetic tsunami waveforms at distant locations." Abstract for IAHS-IAPSO-IASPEI Joint Assembly, SP1S1.03.

Wüthrich, D., Pfister, M., Nistor, I., and Schleiss, A. J. (2018). "Experimental study of tsunami-like waves generated with a vertical release technique on dry and wet beds." *Journal of Waterway, Port, Coastal, and Ocean Engineering*, 144(4), 04018006.

Yeh, H. H., Ghazali, A., and Marton, I. (1989). "Experimental study of bore run-up." *Journal of Fluid Mechanics*, 206, 563–578.

Zahibo, N., Didenkulova, I., Kurkin, A., and Pelinovsky, E. (2008). "Steepness and spectrum of nonlinear deformed shallow water wave." *Ocean Engineering*, 35(1), 47–52.

Zhou, H., Wei, Y., and Titov, V. V. (2012). "Dispersive modeling of the 2009 Samoa tsunami." *Geophysical Research Letters*, 39(16), L16603.

PART 4

Numerical Modelling

Chapter 10

Tsunami Propagation Modelling

10.1 Introduction

A tsunami is a gravity wave phenomenon generated by submarine landslides and earthquakes. Tsunami waves typically have periods of the order of minutes. As such, the waves could be considered waves in shallow water and hence phase velocity of a tsunami can be approximated as $\sqrt{gh}$. With such velocities, the tsunami waves are capable of propagating several thousands of kilometres in a couple of hours. As they close in on the continental shelf and nearshore area, they undergo tremendous amounts of shoaling and diffraction resulting in significant increase in wave height. Thus, tsunamis can cause major destruction along the shoreline in terms of human lives and infrastructure.

A complete spectrum of models is proposed. The reader is expected to gain a complete idea of the models' theoretical and numerical aspects. Therefore, a formal theoretical framework is presented to start with. It is understood that the reader would have acquired the basics of numerical methods elsewhere. Further, certain relevant aspects of numerical modelling are provided. Later on, the theoretical framework and numerical aspects are combined to elaborate modelling of tsunami propagation. Towards the end, the various scale and computation issues are discussed in relation to requirements of different applications ranging from the global scale to regional scale to synoptic scale. To demonstrate the models, various case studies are considered, including a simulation of the 2004 Great Indian Ocean tsunami.

10.2 Equations of Motion

10.2.1 *Navier–Stokes equations*

$$\rho \frac{Du_i}{Dt} = -\frac{\partial p}{\partial x_i} + \mu \frac{\partial^2 u_i}{\partial x_j \partial x_j} + \rho g_i. \tag{10.1}$$

$$\frac{\partial u_i}{\partial x_i} = 0. \tag{10.2}$$

Navier–Stokes equations are the fundamental equations describing the motion of fluids. Equation (10.2) is known as the continuity equation for an incompressible fluid, which simply states that in a control volume, mass must be conserved. The notation used here is the Einstein summation convention where summation is carried out on the repeated index as shown in Eq. (10.2), and i is the repeated index and summation is carried out over this; $u_i(u_1, u_2, u_3)$ is the velocity vector in the x_1, x_2, x_3 orthogonal cartesian directions, and $\frac{D}{Dt}$ (10.1) represents the material derivative which gives us the acceleration of the fluid in an Eulerian framework. The first term on the right-hand side of Eq. (10.1) represents the pressure gradient driving the flow, the second term represents the stress tensor with μ being the molecular diffusivity, and the final term represents the gravitational potential with $g_i(0\,\widehat{x_1} + 0\,\widehat{x_2} + g\,\widehat{x_3})$ being the acceleration due to gravity.

$$\frac{D}{Dt} = \frac{\partial}{\partial t} + u_1 \frac{\partial}{\partial x_1} + u_2 \frac{\partial}{\partial x_2} + u_3 \frac{\partial}{\partial x_3}. \tag{10.3}$$

$$\frac{\partial u_i}{\partial x_i} = \frac{\partial u_1}{\partial x_1} + \frac{\partial u_2}{\partial x_2} + \frac{\partial u_3}{\partial x_3}. \tag{10.4}$$

Turbulence is a state of fluid flow wherein small disturbances to the flow can grow exponentially into large disturbances, forming eddies with characteristic length scales spanning from the large length scales typical of the mean flow all the way down to the small length scales involved in molecular diffusion. To model turbulence, we need to introduce the concept of Reynolds decomposition, wherein we decompose the flow parameter into a time-averaged mean flow parameter and a randomly fluctuating parameter which fluctuates at small time scales about the mean flow. For instance, the instantaneous velocity vector may be decomposed into a time-averaged mean flow velocity vector $\overline{u_i}$ (10.3) with a turbulent fluctuating velocity u_i' oscillating randomly about the mean. The time t_0 in Eq. (10.5) is such that it is smaller than the time scale at which the mean flow varies, but larger than the smaller time scales of the turbulent fluctuations. Here, the time averages of the fluctuating quantities are defined as zero as they are considered to be random processes, i.e., $\overline{u_i'} = 0$.

$$\overline{u_i} = \frac{1}{t_0} \int_0^{t_0} u(t)\,dt. \tag{10.5}$$

When we carry out the time averaging over the Reynolds decomposed Navier–Stokes equations, we obtain what are known as Reynolds Averaged

Navier–Stokes (RANS) equations (10.6).

$$\frac{D\overline{u_i}}{Dt} = -\frac{1}{\rho_o}\frac{\partial \overline{p}}{\partial x_i} + \rho g_i + \frac{\partial}{\partial x_j}\left(\nu\frac{\partial \overline{u_i}}{\partial x_j} - \overline{u_i'u_j'}\right), \tag{10.6}$$

where $\overline{u_i'u_j'}$ is known as the Reynolds stress which acts as an additional stress term due to turbulence on the mean flow. This stress term is symmetric $(\overline{u_i'u_j'} = \overline{u_j'u_i'})$ and introduces six additional unknown terms; however, no additional equations have been obtained. This, in short, is the source of complexity when it comes to modelling turbulence. The manner in which this term is treated numerically depends on what the modeller hopes to achieve and the computational resources available at one's disposal. The simplest models that capture the effect of turbulence are known as algebraic (zero equation) models where the Reynolds stress term is modelled as turbulent diffusivity in addition to molecular diffusivity, with no additional equations, hence the term zero equation. More advanced models include the turbulence energy equation models which are the two equation models such as the $k - \varepsilon$ model which introduces two additional equations describing the turbulent kinetic energy (k) and the rate of dissipation of the turbulent kinetic energy (ε). More advanced models include the Large Eddy Simulation (LES) class of techniques and Direct Numerical Simulations (DNS) with DNS requiring the most intense computational resources.

10.2.1.1 *Boundary condition for RANS*

The boundary conditions will depend on the problem being studied and the turbulence model being employed. For instance, in a $k-\varepsilon$ model, the no-flux condition is typically applied to k at the upper and lower boundaries. For ε, a derivative boundary condition is usually applied at the upper and lower boundaries. The log law of the wall is used to apply derivative boundary conditions to the velocity near the upper and lower boundaries. At the shoreward boundary, a no-flux condition is applied to the velocity, while at the seaward boundary, a Dirichlet boundary condition may be applied using the known sea surface elevation profile for a given tsunami wave.

10.2.2 *System of shallow water equations*

The shallow water equations (SWEs) are derived in the following paragraphs by vertically averaging the continuity equations and the Navier–Stokes equations. The SWE model is shown in Fig. 10.1.

Tsunami

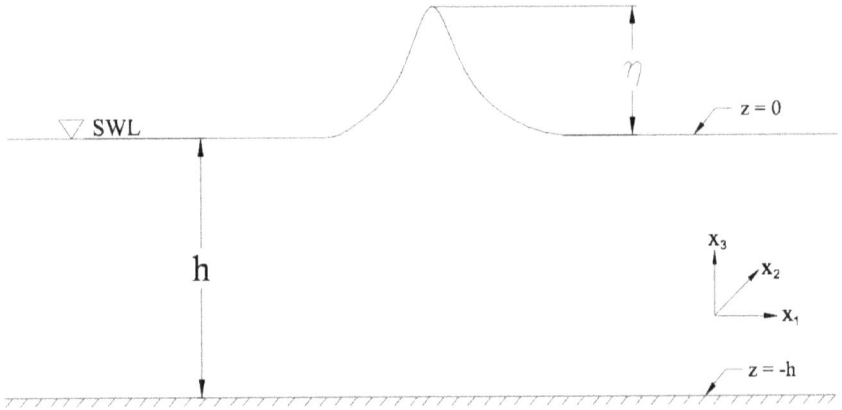

Fig. 10.1   Solitary wave as represented in the SWE model.

We begin by defining the vertically averaged velocity in Eq. (10.7). The continuity equation is vertically integrated in Eq. (10.8).

$$U_i = \frac{1}{(\eta + h)} \int_{-h}^{\eta} u_i \, dx_3. \tag{10.7}$$

$$\int_{-h}^{\eta} \frac{\partial u_i}{\partial x_i} dx_3 = 0. \tag{10.8}$$

Here, $\eta(x_1, x_2)$ is the surface elevation and $h(x_1, x_2)$ is the depth to the bed from the still water level ($x_3 = 0$).

$$\frac{\partial}{\partial x} \int_a^b \phi(x, y) dy = \int_a^b \frac{\partial}{\partial x} \phi(x, y) dy + \phi(x, b(x)) \frac{\partial b(x)}{\partial x} - \phi(x, a(x)) \frac{\partial a(x)}{\partial x}. \tag{10.9}$$

Expanding Eq. (10.8) using the Leibniz integral rule (10.7), we obtain (10.10).

$$\frac{\partial}{\partial x_1} \int_{-h}^{\eta} u_1 dx_3 - u_1(x_1, x_2, \eta) \frac{\partial \eta}{\partial x_1} - u_1(x_1, x_2, -h) \frac{\partial h}{\partial x_1} + \frac{\partial}{\partial x_2}$$

$$\times \int_{-h}^{\eta} u_2 dx_3 - u_2(x_1, x_2, \eta) \frac{\partial \eta}{\partial x_2} - u_2(x_1, x_2, -h) \frac{\partial h}{\partial x_2}$$

$$+ u_3(x_1, x_2, \eta) - u_3(x_1, x_2, -h) = 0. \tag{10.10}$$

We simplify the aforementioned equation using the kinematic free surface boundary condition (KFSBC) (10.11) and the bottom boundary condition

(BBC) (10.12), obtaining the final form in Eq. (10.13). The KFSBC shows that the acceleration of the surface, $\frac{D\eta}{Dt}$, is equal to the acceleration of the fluid particles at the surface. The BBC denotes that the vertical velocity near the bottom is compensated by horizontal divergence which allows the flow to traverse across isobaths.

$$u_3(x_1, x_2, \eta) = \frac{\partial \eta}{\partial t} + u_1(x_1, x_2, \eta)\frac{\partial \eta}{\partial x_1} + u_2(x_1, x_2, \eta)\frac{\partial \eta}{\partial x_2}. \tag{10.11}$$

$$u_3(x_1, x_2, -h) = -u(x_1, x_2, -h)\frac{\partial h}{\partial x_1} - u_2(x_1, x_2, -h)\frac{\partial h}{\partial x_2}. \tag{10.12}$$

$$\frac{\partial \eta}{\partial t} + \frac{\partial[(\eta + h)U_i]}{\partial x_i} = 0. \tag{10.13}$$

Next, we carry out the vertical averaging on the horizontal components (x_1, x_2) of the momentum equations (10.13) and use the hydrostatic assumption (10.14) to represent the pressure term in terms of the gradient of the surface elevation. We finally obtain the form shown in Eq. (10.17).

$$-\frac{1}{\rho}\frac{\partial p}{\partial x} = -g\frac{\partial \eta}{\partial x}. \tag{10.14}$$

$$\frac{DU_i}{Dt} + f \times U_i + g\frac{\partial \eta}{\partial x_i} = \frac{(h+\eta)}{\varrho}\frac{\partial \tau_{ii}}{\partial x_i} + \frac{(h+\eta)}{\varrho}\frac{\partial \tau_{ij}}{\partial x_j}$$

$$+ \frac{1}{\varrho}\tau_{i3}(\eta) - \frac{1}{\varrho}\tau_{i3}(-h). \tag{10.15}$$

To account for the rotation of the Earth, we use the Coriolis parameter, $f(= 2\Omega\sin(\phi)\widehat{x_3})$, where Ω is the rate of rotation of the Earth about its axis, and ϕ is the latitude. The essential assumptions used to derive the SWE are that the pressure is essentially hydrostatic, and the free surface accelerates at the rate of the acceleration of a particle at the surface.

10.2.2.1 Boundary conditions for SWEs

The shoreward boundary may be treated as a no flux boundary, in which case, the model will not be able to represent the inundation and wave run-up. However, to capture inundation, some dry cells may be provided at the shoreward portion of the model's domain which will then be able to capture the inundation.

At the seaward side of the model's domain, the deep water wave velocity is imposed at the boundary as $u = \sqrt{g(\eta + h)}$.

10.2.3 *Modelling inundation: Wetting and drying algorithms*

To model inundation, an algorithm is required to keep track of the wet or dry state of grid points or computational cells. Equation (10.16) is an example of a simple algorithm which keeps track of the distance to which the waterline has moved towards the coast (Hubbert and McInnes, 1999).

$$\Delta X_i^{n+1} = \Delta X_i^n + \Delta t \times \begin{cases} u_{i-1}^n, & n > 0 \\ u_i^n, & n < 0, \end{cases} \qquad (10.16)$$

where ΔX_i is the fractional distance to which the waterline has progressed within the ith cell at the $n + 1^{\text{th}}$ time step, with Δt being the time step. When the waterline has progressed beyond the ith cell ($\Delta X_i > \Delta x$), the cell is considered as being inundated. The same algorithm holds for receding waterline, in which case, when the waterline recedes beyond the ith cell, it would be considered as dry.

10.2.4 *Ray-tracing techniques*

The following are the equations involved in the ray-tracing method spherical coordinates (Satake, 1988).

$$\frac{d\theta}{dT} = \frac{1}{nR}\cos(\zeta). \qquad (10.17)$$

$$\frac{d\phi}{dT} = -\frac{1}{nR\sin(\theta)}\sin(\zeta). \qquad (10.18)$$

$$\frac{d\zeta}{dT} = -\frac{\sin(\zeta)}{n^2 R}\frac{\partial n}{\partial \theta} + \frac{\cos(\zeta)}{n^2 R\sin(\theta)}\frac{\partial n}{\partial \phi} - \frac{1}{nR}\sin(\zeta)\cot(\theta). \qquad (10.19)$$

In Eqs. (10.17)–(10.19), ζ is the ray direction (measured counter-clockwise from South), h is the water depth, $n(=\frac{1}{\sqrt{gh}})$ is the inverse of the wave celerity, R is the radius of the earth, and θ and ϕ are co-latitude (complement of latitude) and longitude, respectively.

The wave ray is normal to the crest of the wave and points in the direction of the wave's propagation at each instant. Ray tracing is computationally less demanding and is therefore a quick way of obtaining important information such as the tsunami wave arrival time, wave path and the convergence or divergence of the wave as it is refracted by the bathymetry of the sea or ocean floor.

10.2.5 Aspects of spatial discretization

Discretization is the process by which a continuous interval is sub-divided into a finite number of points. The computational domain is discretized into a finite number of nodes or elements and the computation is carried out on these discrete nodes or elements which form a computational mesh or grid. When the discretization is applied on the exact governing equation, it is known as the discretized equation with trailing higher order residual terms which will be neglected. These higher-order terms contribute to the discretization errors in the solution. The discretized equation should be consistent with the exact governing equations, in other words, the discretized equations must tend towards the exact governing equations as the discretization step size tend towards zero (Eq. (10.20)). Stability of the discretized equation is a property of the scheme such that the errors do not grow as the scheme marches in time. The consistency and stability of the discretized equations are necessary conditions to ensure convergence to a solution. If the numerical scheme is explicitly being solved, the spatial discretization must satisfy the Courant–Friedrichs–Lewy (CFL) condition which is a necessary condition to ensure stability of the numerical scheme. The CFL criterion places a limit on the time stepping and the spatial discretization. The stability criterion may be found for each type of numerical scheme using various techniques such as the von Neumann stability analysis.

To ensure stability in the case of advection terms in hyperbolic equations, the spatial discretization should be such that the grid points from the upstream side of the flow are selected, this is known as upwinding.

$$\lim_{(\Delta x, \Delta t) \to 0} (\text{Discretized Equations}) = (\text{Exact Governing Equations}).$$
$$(10.20)$$

To obtain lower errors near regions of strong gradients or regions of interest, it is possible to increase the resolution of the computational grid locally. Such a non-uniform mesh then needs to be transformed onto a computational coordinate system in which the grid remains uniform. Once the computation is carried out, the solution is once again transformed onto the non-uniform mesh.

10.2.6 Time-stepping aspects of spatial discretization

Time-stepping scheme must be selected such that the discretized equation is consistent and stable. Consistency is straightforward to analyse; however, stability analysis techniques such as the von Neumann stability analysis can become quite complex to solve. The stability criterion places an upper limit on the time step size. However, a stable scheme may still have high discretization errors which arise due to the residual higher-order terms. These may be reduced by further reducing the time step size. The time scales of the physical phenomenon being studied should also be considered while selecting the time step size. If the time step exceeds the time period of the flow phenomenon that one wishes to capture, the numerical solution may not represent the phenomenon or may show false solutions that are an artefact of the numerical scheme. The time step must be lesser than or equal to the Nyquist frequency (which is half the frequency of the phenomenon being studied).

10.3   Case Studies and Solutions

10.3.1   *Shallow water equations*

Siva and Behera (2016) carried out numerical experiments using a shallow water model to study the effect of varying continental slopes on the approaching tsunami wave. Figure 10.2 shows the typical "N" wave propagating shoreward and the modification that it undergoes as it encounters the continental slope. The slopes used ranged from 1:0.1 to 1:75 and were selected based on the typical continental slopes found around the Indian Coast. The study found that, in general, as the slope becomes gentler, the ratio of height of the wave on the shelf to the height of the wave off shelf

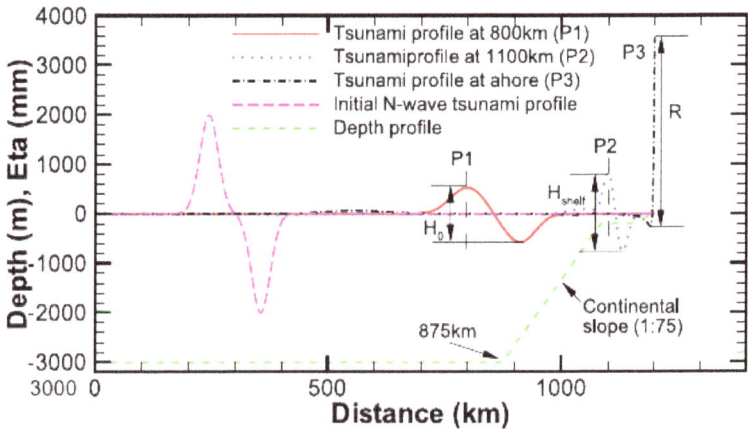

Source: Siva and Behera (2016).

Fig. 10.2   Tsunami of "N" wave initial profile propagating and undergoing modification as it encounters the shelf slope.

(H_s/H_o) becomes greater. In other words, the steeper slopes cause greater energy dissipation of the approaching tsunami wave resulting in a lower wave height on the shelf. However, this effect was most pronounced when the shelf depth was low, and became less pronounced with increasing shelf depth. The experiments were carried out for shelf depths of 50 m, 75 m, 100 m and 200 m and it was found that H_s/H_o increases by 144% with decrease in slope for a shelf depth of 50 m and the corresponding increase in H_s/H_o for a shelf depth of 200 m was found to be 43%.

10.3.2   *Finite Element Models (FEMs)*

Behera *et al.* (2009) used an Unstructured Explicit FEM based on the model developed by Chitra *et al.* (1996) to simulate the propagation of the Indian Ocean Tsunami of December 2004. Figure 10.3 shows the wave run-up heights computed in the model and it compares favourably with the observations at various points along the coast of Tamil Nadu. The model

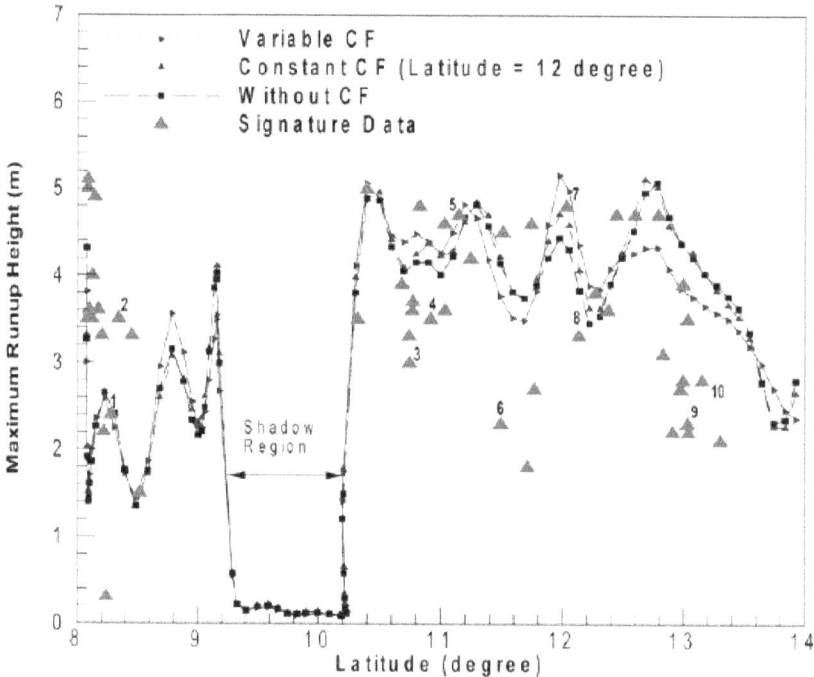

Source: Manasa *et al.* (2009).

Fig. 10.3   Simulated wave run-up heights of the Indian Ocean Tsunami of December 2004 along the east coast of India.

also captured the effect of the Sri Lankan island functioning as a barrier to
the propagation of the wave with the model predicting lower run-up heights
in the shadow region along the coast of Tamil Nadu.

The finite element methods based on FNPT was used to study the transmission and reflection of tsunami waves used FEMs to study the transmission and reflection of tsunami waves propagating on (i) a bed with a step,
(ii) on a bed with various slopes, and (iii) on a bed with a vertical wall. This
model is able to represent the splitting of the tsunami wave due to nonlin-
earities. This study too shows that the reflection at the bed slope reduces
with reducing bed slopes. Figure 10.4 shows the highest η of reflected waves

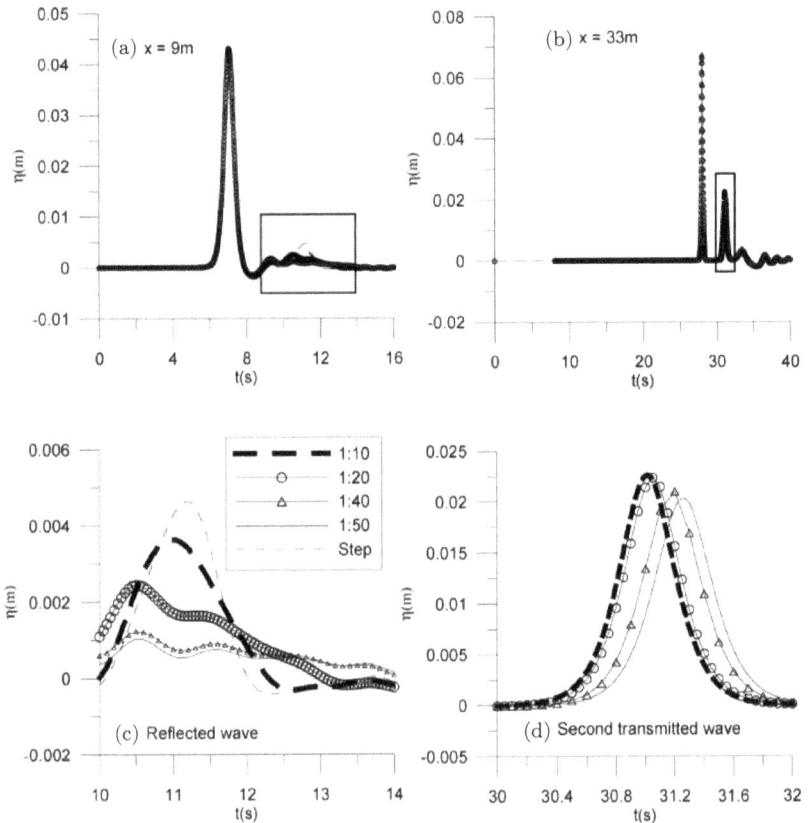

Source: Sriram *et al.* (2006).

Fig. 10.4   Solitary wave undergoing modification as it traverses a gentle slope.
(a) is at a point upstream of the slope. (b) is at a point downstream of the slope.
(c) is a zoomed in view of the window shown in (a). (d) is the zoomed in view of
the window shown in (b).

for the case with an abrupt step, followed by the steepest slope of 1:10 with the reflected η reducing as the slope becomes gentler.

10.3.3 Ray tracing of tsunami waves

Satake (1988) carried out a numerical ray tracing of tsunami waves in the Pacific Ocean and the Japan Sea. Ray tracing is computationally cheap and allows for very quick computation of wave arrival times, wave travel paths, and the convergence or divergence of the wave energy as the wave refracts over the bathymetry. It is suitable for identifying the coastal regions that could be most affected by the tsunami waves which could help inform where the computationally more intensive higher resolution FEM models or RANS models may then be applied to obtain the wave loads, run-up and inundation at these specific locations. Satake (1988) further showed that in the Pacific Ocean, the dominant effect is that of convergence or divergence due to the bathymetry; however, in the Japan Sea, the waves were also affected by the presence of an oceanic rise where strong refraction occurred.

Prasad Kumar *et al.* (2006) presented a computationally efficient algorithm to compute the travel time of tsunami waves from the point of genesis to all densely populated points along the coastline surrounding the Indian Ocean. Figure 10.5 shows the wave travel time computed for the December 2004 Indian Ocean Tsunami event. The algorithm consists of time stepping across discretized nodes using the long wave velocity ($\sqrt{(gh_i)}$), where h_i is the depth at the ith node. This algorithm may function as an effective early warning system as it may be computed within minutes to seconds of the occurrence of an earthquake that could cause a tsunami.

10.3.4 RANS modelling of inundation due to tsunami wave

Choi *et al.* (2012) applied a RANS model to capture the full spectrum of the phenomena associated with tsunami waves, right from the propagation of the wave in deep ocean to the inundation caused on the shore. The study used equivalent resistance coefficients which suitably modifies the Manning's roughness coefficient to take into account the drag effect exerted by the buildings on the flow by considering factors such as the amount of built up area and the shape of the buildings. The equivalent resistance coefficient was computed using different analytical and semi-analytical formulae

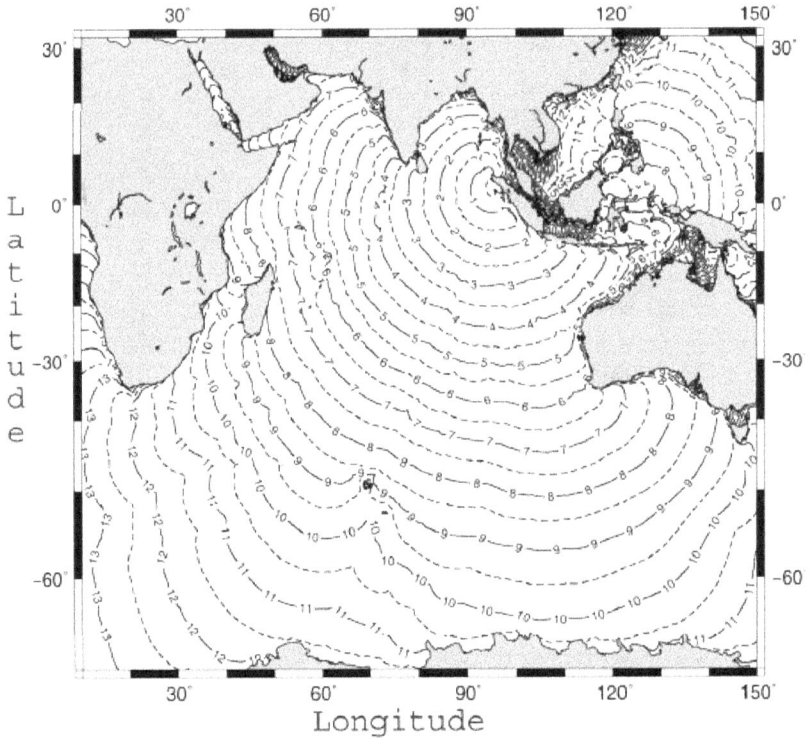

Source: Prasad Kumar *et al.* (2006).

Fig. 10.5   Tsunami wave travel time for the December 2004 Indian Ocean tsunami event.

and the same was compared with laboratory experiments and numerical experiments. The numerical model used a $k - \varepsilon$ turbulence scheme and buildings in the urban built-up area of the shore were modelled as rigid square piers. Once the equivalent resistance coefficient was tuned, it was used in the numerical model to replicate the effects on the Imwon harbour of South Korea due to the 1983 tsunami in the Sea of Japan. Figure 10.6 shows that the tuned resistance coefficient predicts inundation better than models using a constant roughness coefficient. The highest flood level predicted by the model using the equivalent resistance coefficient was closer to the observed high flood line than the model using a constant roughness coefficient.

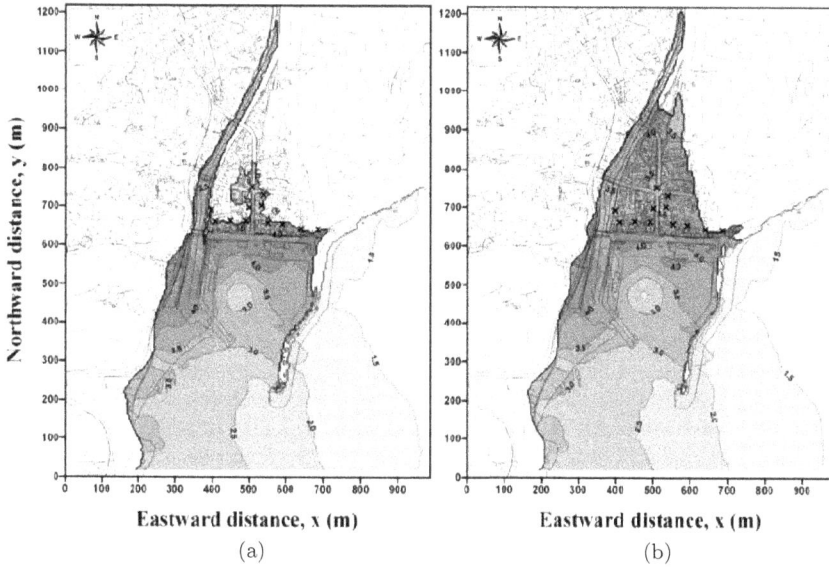

Source: Choi et al. (2012).

Fig. 10.6   Inundation of Imwon harbour, South Korea. (a) Results of model using equivalent resistance coefficient. (b) Results of model using constant roughness coefficient. Crosses indicate the observed maximum extent of inundation.

10.4   Summary

The aspects of numerical modelling of SWEs, boundary conditions, and methods of solving the equation to understand the propagation of a tsunami from its source have been highlighted in this chapter. The wetting and drying conditions have also been considered. The validation of the numerical model with signature studies carried out after the 2004 tsunami is reported in this chapter.

References

Behera, M. R., Murali, K., Sannasiraj, S. A., and Sundar, V. (2009). "Simulation and prediction of runup heights due to great Indian Ocean tsunami," in *Advances in Water Resources and Hydraulic Engineering*, Springer, Berlin, 1244–1249.

Chitra, K., Murali, K., Mahadevan, R., and Aswathanarayana, P. A. (1996). "Simulation of storm surges using an Explicit FEM." International Conference in Ocean Engineering, IIT Madras, India.

Choi, J., Kwon, K. K., and Yoon, S. B. (2012). "Tsunami inundation simulation of a built-up area using equivalent resistance coefficient." *Coastal Engineering Journal*, 54(2), 1250015-1–1250015-25.

Hubbert, G. D., and McInnes, K. L. (1999). "A storm surge inundation model for coastal planning and impact studies." *Journal of Coastal Research*, 168–185.

Manasa R. B., Murali K., Sannasiraj S., and Sundar V. (2009). "Simulation and prediction of runup heights due to great Indian Ocean tsunami." In: *Advances in Water Resources and Hydraulic Engineering*.

Prasad Kumar, B., Rajesh Kumar, R., Dube, S. K., Murty, T., Gangopadhyay, A., Chaudhuri, A., and Rao, A. D. (2006). "Tsunami travel time computation and skill assessment for the 26 December 2004 event in the Indian Ocean." *Coastal Engineering Journal*, 48(02), 147–166.

Satake, K. (1988). "Effects of bathymetry on tsunami propagation: Application of ray tracing to tsunamis." *Pure and Applied Geophysics*, 126(1), 27–36.

Siva, M., and Behera, M. R. (2016). "Effect of continental slope on N-wave type tsunami run-up." *The International Journal of Ocean and Climate Systems*, 7(2), 47–54.

Sriram, V., Sannasiraj S. A., and Sundar V. (2006). "NWF: Propagation of tsunami and its interaction with continental shelf and vertical wall." *Marine Geodesy*, 29(3), 201–221.

Chapter 11

Tsunami Evolution and Run-up

11.1 Introduction

In this chapter, the modelling aspects of the tsunami evolution and run-up characteristics using fully nonlinear potential flow theory and viscous models based on particle method using hybrid coupling will be discussed. Based on the past events, tsunamis approaching the shore may broadly be classified as (a) a series of split waves, (b) non-breaking waves that act as a rapidly rising tide and (c) a large, turbulent wall-like wave. The modelling aspects of the series of split waves and the dynamics of strongly nonlinear waves including breaking, formation of shock waves and modelling the run-up aspects will also be discussed. Further, the chapter discusses the limitation of the one-parameter solitary wave for such a complex multi-parameter phenomenon like a tsunami using the numerical models.

11.1.1 *General*

Tsunamis are extreme waves that are generated due to an earthquake in the ocean bottom or any other disturbances in the ocean surface/bottom. As they arrive near the shore, they are characterized by very high waves with a long period, whereas in the deep ocean they are characterized by a long wave period with small amplitude. There have been a significant number of tsunami events recorded in almost all parts of the world. The previous encounters with tsunamis have helped us in understanding the havoc they cause to the infrastructure and human civilization. One of the important factors that needs to be known is the run-up and inundation. In the previous chapter, it was clearly brought out that a tsunami cannot be generated in the flumes. So, the solitary waves were used to model the tsunami waves in the laboratory environment (Synolakis, 1987), and run-up from the solitary waves is widely adopted. However, the solitary waves are characterized by a single positive pulse and this in turn results in a wave with no trough. It was observed during the tsunami events that the shoreline tends to recede before the onset of the wave. In many of the reported cases, this phenomenon had caused an increase in the number of fatalities. The study done by Madsen *et al.* (2008) had suggested against the use of solitary wave theory as a tsunami model. Compared to a solitary wave, N-wave has a distinctive trough and crest. Based on the relative shape of the crest and trough, the N-wave can be symmetric or asymmetric. A leading trough N-wave model can exactly model the receding shoreline phenomenon which was evident during the field observations. It was observed during our previous study (Manoj *et al.*, 2017) that the breaking characteristics and effective run-up of both the solitary and N-wave models were quite different even though they had the same wave height.

Experiments in the model scale, analytical formulations and numerical models are some of the most reliable tools when it comes to the study of wave evolution and run-up during extreme wave events. An extreme wave event such as a tsunami traces its origin to deeper waters and usually travels for very long distances. Schimmels *et al.* (2016) and Sriram *et al.* (2016) investigated the evolution and run-up of the tsunami in a large-scale experimental facility. Further, the run-up and inundation of the generated waves were studied using the analytical model developed by Didenkulova (2009). In the aforementioned study, it was observed that the run-up on a mild slope (>1:18 with respect to the study) was difficult to estimate due to the phenomenon of wave breaking during the run-up, which was observed as one of the limitations of the analytical model.

The important aspect in the study of tsunami waves is the run-up associated with the wave. The estimation of run-up along the beach helps in protecting the people and infrastructure. Empirical models have been proposed by the researchers to estimate the run-up of tsunami-like waves. The estimations were based on rigorous model studies performed in the laboratory scale using solitary, Cnoidal and N-wave models. In many of the cases, the results were supported by numerical studies based on the shallow water equations, potential flow theory and Navier–Stokes equation model.

In this chapter, the experimental data from GWK Hannover will be used as an input to the numerical model to simulate the breaking waves over the slope and simulate the run-up characteristics. The numerical method is based on Sriram *et al.* (2014). The method combines mesh-based finite element method (FEM) and the particle-based meshless local Petrov–Galerkin method.

11.2   Formulation of the Problem

11.2.1   Mathematical formulation

The two-dimensional fluid motion is defined with respect to the fixed Cartesian coordinate system, Oxz, with the z-axis positive upwards. The water depth d is assumed to be a constant unless mentioned otherwise and L is the length of the tank. The left region of the computational domain is solved using the FNPT solver, whereas on the right region, the problem will be solved using the NS solver, wherein the wave breaking or any structure interaction takes place. The sketch of the computational domain is shown in Fig. 11.1.

11.2.2   Fully nonlinear potential theory and finite element method

In the domain for potential flow, the fluid is assumed to be incompressible and viscous forces are neglected. This simplified flow problem is defined by Laplace's equation about velocity potential $\Phi(x, z)$ given by the following:

$$\nabla^2 \Phi = 0. \tag{11.1}$$

A potential flow in a sub-domain with a wave maker at one end and the nonlinear free surface boundary condition is considered. The prescribed von Neumann and Dirichlet boundary conditions are applied on the boundaries. Considering the rigid flume bottom as flat, one can write the following:

$$\frac{\partial \Phi}{\partial z} = 0 \quad \text{at } z = -d \text{ on } \Gamma_B. \tag{11.2}$$

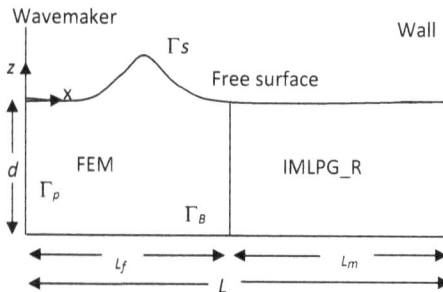

Fig. 11.1   Computational domain and coordinate system.

Motion of the wave paddle at the left end can be enforced by the following:

$$\frac{\partial \Phi}{\partial x} = \dot{x}_p(t) \quad \text{at } x = x_p(t) \text{ on } \Gamma_p, \tag{11.3}$$

where $x_p(t)$ is the time history of wave paddle motion at $x = 0$. The non-linear dynamic free-surface condition in a Lagrangian form is given by the following:

$$\frac{dx}{dt} = \frac{\partial \Phi}{\partial x}, \tag{11.4}$$

$$\frac{dz}{dt} = \frac{\partial \Phi}{\partial z}, \tag{11.5}$$

$$\frac{d\Phi}{dt} = -P - gz + \frac{1}{2}|\nabla \Phi|^2, \tag{11.6}$$

where $P = 0$ and $z = \eta$ on the free surface.

The solution for the aforementioned initial boundary value problem (IBVP) is solved using a finite element scheme. Formulating the governing Laplace equation constrained with the associated boundary conditions in Eqs. ((11.2)–(11.6)) leads to the following finite element systems of equation:

$$\int_\Omega \nabla N_i \sum_{j=1}^m \phi_j \nabla N_j d\Omega|_{j,i\notin\Gamma_s} = -\int_{\Gamma_p} N_i \dot{x}_p(t) d\Gamma$$

$$-\int_\Omega \nabla N_i \sum_{j=1}^m \phi_j \nabla N_j d\Omega|_{j\in\Gamma_s, i\notin\Gamma_s}, \tag{11.7}$$

where "m" is the total number of nodes in the domain, Γ_s is the free surface boundary and the velocity potential inside an element $\Phi(x, z)$ can be expressed in terms of its nodal potentials ϕ_j as follows:

$$\Phi(x, z) = \sum_{j=1}^n \phi_j N_j(x, z).$$

Herein, N_j is the shape function and n is the number of nodes. Linear triangular elements with structured mesh are used in this chapter, though the method can be used for any kind of mesh (Ma et al., 2009). At the start of the simulation ($t = 0$) of the wave generation problem, the free surface elevation $\eta(x, 0)$ and velocity potential, $\Phi(x, z, 0)$ are assumed to be zero. Velocities are evaluated based on least squares method (Sriram, 2008) after the solution for the potential is found. The procedure has been described in our previous publications (e.g., Sriram, 2008; Sriram et al., 2006, 2010) and is summarized in the following sections for completeness.

Let us assume that the velocity potential and free surface elevation are known at the previous time step. Then the following two-step time integration (Runge–Kutta second-order method) is used to find the new position and velocity potential for the next time step by using Eqs. ((11.4)–(11.7)).

(1) At time $t_n = t^+$, solve the BVP in Eq. (11.7) based on the previous time-step velocity potential at Dirichlet boundary and find its derivatives in space and time (using Eqs. (11.4)–(11.6)) and compute,

$$k_{1a} = (\partial\phi^n/\partial x, \partial\phi^n/\partial z), \quad k_{1b} = d\phi^n/dt. \tag{11.8a}$$

(2) Find the intermediate values of coordinates and velocity potential on the Dirichlet boundary condition using explicit Euler Formulae:

$$\vec{r}^{+\,(n+1)} = \vec{r}^n + k_{1a}\,dt, \quad \phi^{+\,(n+1)} = \phi^n + k_{1b}\,dt. \tag{11.8b}$$

(3) Solve the BVP again (Eq. (11.7)) for the intermediate values and find its derivatives in space and time.

$$k_{2a} = (\partial\phi^{+n+1}/\partial x, \partial\phi^{+n+1}/\partial z), \quad k_{2b} = d\phi^{+n+1}/dt. \tag{11.8c}$$

(4) Calculate the final values of the coordinates and velocity potential for the next time step based on the average of the values found in step (1) and (3).

$$\vec{r}^{(n+1)} = \vec{r}^n + (0.5k_{1a} + 0.5k_{2a})\,dt,$$

$$\phi^{(n+1)} = \phi^n + (0.5k_{1b} + 0.5k_{2b})\,dt. \tag{11.8d}$$

11.3   Navier–Stokes Solver (NS/IMLPG_R)

The Navier–Stokes equation (referred to as NS equation) and continuity equations together with proper boundary conditions are considered as follows:

$$\frac{D\vec{u}}{Dt} = -\frac{1}{\rho}\nabla p + \vec{g} + \nu\nabla^2\vec{u}, \tag{11.9}$$

$$\nabla \cdot \vec{u} = 0, \tag{11.10}$$

where g is the gravitational acceleration, $\vec{u}$ is the fluid velocity vector, p is the pressure and ρ is the density of the fluid. The Lagrangian forms of the kinematic and dynamic conditions on the free surface are given as follows:

$$\frac{D\vec{r}}{Dt} = \vec{u}. \tag{11.11}$$

$$p = 0, \tag{11.12}$$

where $\vec{r}$ is the position vector. On the rigid boundary surface, the following boundary conditions are satisfied.

$$\vec{u} \cdot \vec{n} = \vec{U} \cdot \vec{n} \tag{11.13}$$

and

$$\vec{n} \cdot \nabla p = \rho(\vec{n} \cdot \vec{g} - \vec{n} \cdot \dot{\vec{U}}), \tag{11.14}$$

where $\vec{n}$ is the unit normal vector of the rigid boundary, $\vec{U}$ and $\dot{\vec{U}}$ are its velocity and acceleration, respectively. It is noted that the slip boundary condition is applied on the bottom and so the effects of boundary layer or shear stresses there are not considered. In the cases where the effects of the boundary layer or shear stresses on the bottom are more significant, they should be investigated carefully.

If one knows the velocity, pressure and the position of the particles at nth time step ($t = t^n$), then in the IMLPG_R algorithm the following procedure is used to find the variables at $(n+1)$th time step.

(1) Calculate the intermediate velocity ($\vec{u}^*$) and position ($\vec{r}$) of particles using

$$\vec{u}^* = \vec{u}^n + \vec{g}dt + \nu\nabla^2\vec{u}^n dt, \tag{11.15}$$

$$\vec{r}^* = \vec{r}^n + \vec{u}^* dt, \tag{11.16}$$

(2) Evaluate the pressure p^{n+1} using the following semi-implicit equation (Sriram and Ma, 2012).

$$\nabla^2 p^{n+1} = \frac{\rho}{dt}\nabla\vec{u}^*. \tag{11.17}$$

(3a) Calculate the velocity at all the particles using the pressure gradient.

$$\vec{u}^{**} = -\frac{dt}{\rho}\nabla p^{n+1}. \tag{11.18}$$

$$\vec{u}^{n+1} = \vec{u}^* + \vec{u}^{**} = \vec{u}^* - \frac{dt}{\rho}\nabla p^{n+1}. \tag{11.19}$$

(3b) Update the position of the inner particles (that are not on the free surface or on boundary) using the following:

$$\hat{\vec{u}}^{n+1} = \vec{u}^* - \frac{dt}{\rho}\nabla\hat{p}^{n+1}. \tag{11.20}$$

$$\hat{\vec{r}}^{n+1} = \vec{r}^n + \hat{\vec{u}}^{n+1}dt, \tag{11.21}$$

where $\hat{}$ in the aforementioned equations corresponds to the estimation based on the modified pressure gradient (Koshizuka and Oka, 1996; Sriram and Ma, 2012).

(3c) Update the positions of free surface and boundary particles using the following:

$$\vec{r}^{n+1} = \vec{r}^n + \vec{u}^{n+1}dt. \tag{11.22}$$

(4) Go to the next time step.

The aforesaid procedure is adopted in the improved meshless local Petrov–Galerkin method (IMLPG_R) (Sriram and Ma, 2012; Khayyer and Gotoh, 2009). As it can be seen, the key task of this procedure is to solve Eq. (11.9) in order to evaluate the pressure. Numerical techniques, like, finite element and finite difference methods may be adopted to solve these equations. In our work, the MLPG_R method is used. The details of the MLPG_R formulation and other techniques have been discussed by Zhou *et al.* (2009, 2010) and Sriram and Ma (2012). Only the basic principle is given here and more details can be found in the cited papers. In this method, the fluid domain is represented by a number of particles. The governing equation, Eq. (11.17), is transferred into the following weak form by integrating it over a circular sub-domain surrounding each node.

$$\int_{\partial\Omega_I} \vec{n} \cdot (p\nabla\varphi)dS - p = \int_{\Omega_I} \frac{\rho}{dt}\vec{u}^* \cdot \nabla\varphi \, d\Omega, \tag{11.23}$$

where $\vec{n}$ is the unit normal vector pointing outside of the sub-domain, $\varphi = \frac{1}{2\pi}\ln(r/R_I)$ is the solution for Rankine source in an unbounded 2D domain with r being the distance between a concerned point and the centre of the local sub-domain Ω_I and with R_I being the radius of Ω_I. The major difference of this equation from Eq. (11.17) is that it does not include any derivative of unknown functions, while Eq. (11.17) contains the second-order derivatives of unknown pressure and gradient of velocity. Approximation to the unknown functions in Eq. (11.23) does not require them to have any continuous derivatives, while approximation to the unknown functions in Eq. (11.17) requires them to have finite or at least integrable second-order derivatives. Therefore, use of Eq. (11.23) for further discretization has a great numerical advantage over use of Eq. (11.17) directly. One of differences between our MLPG_R method and MPS (or incompressible SPH) method (the latter discretising Eq. (11.17) directly) lies in use of the different pressure governing equations for further discretization, as indicated earlier.

Ma and Zhou (2009) have detailed the method to discretize Eq. (11.23), in which the pressure on the left-hand side is interpolated by a moving least square (MLS) method and the integration on the right-hand side is evaluated by a semi-analytical technique. The derivation will not be repeated here but only the final equation is given as follows:

$$\mathbf{A.P} = \mathbf{B} \tag{11.24}$$

where

$$p(\vec{x}) \approx \sum_{j=1}^{N} \Phi_j(\vec{x})\hat{p}_j$$

$$A_{ij} = \begin{cases} \int_{\partial\Omega} \Phi_j(\vec{x}) \cdot \vec{n} \cdot \nabla\varphi ds - \Phi(\vec{x}) & \textit{for inner nodes} \\ \vec{n} \cdot \nabla\Phi(\vec{x}) & \textit{for solid boundary nodes} \end{cases}$$

$$B_i = \begin{cases} \int_{\Omega} \dfrac{\rho}{dt}\vec{u}^* \cdot \nabla\varphi d\Omega & \textit{for inner nodes} \\ \dfrac{\rho}{dt}\vec{n} \cdot (\vec{u}^* - \vec{U}^{n+1}) & \textit{for solid boundary nodes,} \end{cases}$$

where $\Phi_j(\vec{x})$ is the shape function which is evaluated by using the MLS method as described by Ma and Zhou (2009) and Ma (2005). The water particles discretizing the fluid domain are separated into two groups: those not on the free surface (referred to as inner particles) and those on the free surface (referred as free-surface particles). The free surface particles are identified at the beginning of the calculation and then they are identified at every time step by using the mixed particle number density and auxiliary function (MPNDAF) method developed by Ma and Zhou (2009) when applied to modelling violent waves.

Once the solution for the pressure is found, the gradient of pressure needs to be estimated in order to update the velocity and the positions of the water particles. The estimation of pressure gradient is made by using the simplified finite difference scheme (SFDI) (Ma, 2008).

11.4 Coupling Methodology in Space Domain

As it has been pointed out, there are two types of methods coupling potential flow models with viscous flow models. One is weak coupling and the other is strong coupling. The weak coupling is faster as it needs to only transfer information in one direction and holds good for some problems. However, when carrying out the simulations for a long-time wave-structure interaction, one needs to resort to a strong coupling procedure in which the information from both the solvers is exchanged, and so solutions from both the solvers affect each other. In this chapter, the results from the strong coupling procedure are reported. In general, the strong coupling needs to couple the models both in space and time domains. For the coupling in the space domain, the following three methods have been found to be employed (Grilli, 2008; Sueyoshi *et al.*, 2007; Sitanggang, 2008):

(1) Fixed boundary interface.
(2) Moving boundary interface.
(3) Fixed overlapping zone interface.

Options (1) and (3), where the interfaces are fixed, hold good if both the solvers are based on the Eulerian formulation and fixed meshes, for example, Grilli (2008) and Sitanggang (2008). Option (3) was also used for SPH-one-way coupling (Kassiotis *et al.*, 2011), but they need to add or delete the particles at every time steps in the overlapping zone. In this option, the conservation of mass is questionable. The second approach is potentially suitable for the Lagrangian formulation. However, if adopting this approach, one may have difficulty in satisfying the conditions on the moving interface. That is because the conditions and variables to be solved are different on the different sides of the interface as the fluid flow is potential in one region and viscous in the other region.

It is clear from the aforesaid discussions that none of the said three approaches for the coupling are ideal for our hybrid model because the FNTP-based finite element method and the NS-based IMLPG_R to be used in this chapter are either based on the Eulerian–Lagrangian formulation or pure Lagrangian formulation. To overcome the difficulty, a new approach is developed and reported in Sriram *et al.* (2014), which is called moving overlapping zone interface. It is similar to Option (3) in the sense that there is an overlapping zone. The difference between the new one and the fixed one is that the overlapping zone is allowed to move and its boundary is allowed to vary. More detailed discussions about this are given in the following sections.

11.4.1 *Moving overlapping zone*

As it has been described previously, the whole computational domain is split
into two sub-domains, one where the FNPT-based finite element method
applies and the other where the NS-based IMLPG_R method is employed.
For simplicity, the two regions will be denoted by FEM and IMLPG_R,
respectively. In the FEM region, there will be no wave breaking or over-
turning. The technique of *moving overlapping zone* will be adopted to couple
the two solvers. It is further depicted in Fig. 11.2 that the overlapping zone
does not only move but its shape also varies during simulation. The FNPT

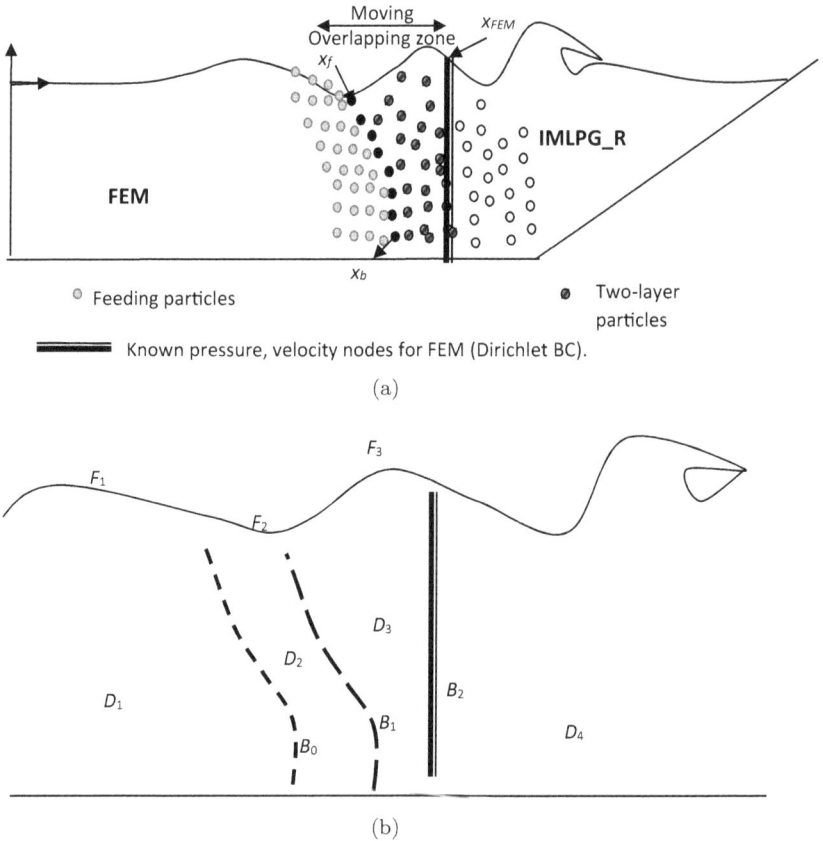

(a)

(b)

Fig. 11.2 Illustration of coupling technique — moving overlapping zone.
(a) Details of overlapping zone, two-layer particles and coupling details. (b)
Details of the computation domain and boundary descriptions used in the hybrid
model (D: Domain, B: Boundary, F: Free surface boundary).

model is applied in the region up to B_2 with B_2 as its boundary, while
the NS model is applied in the region after B_1 with B_1 as its boundary.
B_2 moves with a water particle on the free surface and is kept as straight.
B_1 varies with particles on it and so can be curved. The solution from
the FNPT model is feed to the NS model through B_1 while the solution
from the NS model is the feedback to the FNPT model through B_2. The
following three issues are needed to be addressed.

(1) Treatment of FEM boundary (B_2) in the IMLPG_R domain.
(2) Treatment of IMLPG_R boundary (B_1) in the FEM domain.
(3) Treatment of the velocity in the overlapping zone.

11.4.1.1   *Treatment of FEM boundary in IMLPG_R domain*

The information on the FEM boundary B_2 must be obtained from the solution of the IMLPG_R method. The difficulty is that the unknown variable
to be solved in the FEM is the velocity potential and so the boundary
value on B_2 should be given in terms of the velocity potential, whereas the
IMLPG_R method can only provide the solutions for velocity and pressure.
In addition to this, the nodes for the FEM are generally not coincident
with the particles for the IMLPG_R method. If B_2 follows the particles
used for the IMLPG_R method, large deformation of elements for the FEM
may occur and so cause numerical problems. To avoid this, it is proposed
that the B_2 boundary be straight and vertical and move horizontally with
a particle on the free surface, similar to that for dealing with the radiation
boundary in (Ma *et al.*, 2001; Yan, 2006).

The positions of nodes on B_2 for the FEM calculation and the value of
velocity potential on them is estimated by the following equations:

$$\frac{dx}{dt} = (u)_{\bar{x}_f \text{ IMLPG\_R}} \quad \frac{dz}{dt} = (w)_{\text{IMLPG\_R}}, \qquad (11.25a)$$

$$\left.\frac{\partial \Phi}{dt}\right|_{B_2} = -\left(P + \frac{1}{2}|\vec{u}|^2 - u \cdot u_f\right)_{\text{IMLPG\_R}} - gz, \qquad (11.25b)$$

where P is the dynamic pressure, the subscript IMLPG_R refers the variables calculated by using the solution of the IMLPG_R method, x_f corresponds to the free surface node on B_2 and u_f corresponds to the velocity
of the nodes used for the FEM and calculated by using $((\vec{x}^n - \vec{x}^{n-1})/dt)$,
with n referring to current time step. It is noted that the reason for the
term with u_f to be considered is because the nodes on B_2 have normally
different velocity from $(u)_{\bar{x}_f \text{ IMLPG\_R}}$.

As indicated earlier, the nodes for the FEM generally do not coincide with the particles for the IMLPG_R. As a result, the pressure and velocity for Eq. (11.25) need to be interpolated from the IMLPG_R solution. The interpolation scheme used for this purpose is based on the MLS scheme as described by Ma (2005). One point is noted here. If the total pressure from IMLPG_R method would have been used in Eq. (11.25), the oscillation in pressure is severe for a long time computation. This problem was also noticed and reported by Sueyoshi *et al.* (2007) when coupling the BEM method with the MPS method in their work. This is why only dynamic pressure is estimated from the IMLPG_R solution but the static pressure is added explicitly in Eq. (11.25).

11.4.1.2   *Treatment of IMLPG_R boundary in the FEM domain*

For the IMLPG_R boundary (B_1), similar issues to those discussed for B_2 previously need to be addressed, that is, how to move the particles and how to determine the values of physical variables on B_1, though the way to deal with them must be different. As for moving particles on B_1, the consideration should be given that the IMLPG_R method is based on the fully Lagrangian formulation, and so the position of a particle is determined by the fluid velocity. In water waves, the fluid velocity on the free surface is generally much larger than that at the bottom, though the difference becomes smaller in shallow water cases. In our research, tests have been carried out to use the straight line for B_1 and move it in the similar way as for B_2. When doing so, one needs to add and delete particles for the IMLPG_R method as done by Kassiotis *et al.* (2011) for dealing with the inlet and outlet boundaries of their SPH method. It is found that large error can occur due to lack of mass conservation. We instead propose another approach here, that is, the particles on B_1 are always kept on it, and so the shape of B_1 is determined by the position of the particles, whose velocity is determined by interpolating the velocity based on the solution in the FEM domain using the method detailed in Sriram (2008). As the particles' velocity is equal to fluid velocity, the B_1 generally becomes curved during simulation even though it may be selected as a straight line at start. The curved line will always be single valued as waves will not be overturned (so not broken) in the FEM region before B_2 as indicated earlier. With this approach, the particles for the IMLPG_R method will never move out of the region for it, and so there does not exist the issues of mass non-conservation. However, it is noted here that the velocity of particles on B_1 determined in this way may not be continuous with the velocity at the particles in the IMLPG_R region. This issue will be addressed in the next section.

As for determining the values of physical variables on B_1, it is necessary to clarify what is needed in the process to find the solution in the IMLPG_R region. Firstly, the boundary value of pressure on B_1 is required in order to solve Eq. (11.23) or (11.24). Secondly, the pressure gradient near B_1 is necessary in order to evaluate the velocity of particles within the IMLPG_R domain.

The pressure at the particles on the IMLPG_R boundary B_1 is determined by Eq. (11.6), i.e.,

$$P|_{B_1} = \left(\frac{d\Phi}{dt} - \frac{1}{2}\vec{u} \cdot \vec{u} \right)_{\text{FEM}} - gz, \tag{11.26}$$

where the subscript FEM indicates that the values are estimated by using the solution in the FEM domain. Again the hydrostatic pressure is directly evaluated by the position of the particle concerned. There are two issues associated with the evaluation in Eq. (11.26). The first one is how to determine the time derivative of the velocity potential. Although it may be theoretically estimated by a finite difference scheme from the solution of two successive time instants in the FEM domain, it has been well known that using this way may cause instability (Kim *et al.*, 1999). In this chapter, the time derivative of the velocity potential is found by solving a boundary value problem for $\frac{d\Phi}{dt}$, formed by Laplace equation and von Neumann boundary conditions on wave paddle and bottom, and Dirichlet boundary conditions on the free surface and B_2, similar in some way to those in Eqs. (11.1)–(11.6) for finding the velocity potential. More details about this can be found in Ma *et al.* (2001), Ma and Yan (2006) and Sriram (2008). The second issue is that solutions for the velocity potential and its derivatives in the FEM domain are given at the nodes for the FEM method but the pressure needs to be evaluated at the particles for the IMLPG_R method. They are not at the same positions as indicated previously. Because of this, the velocity and the values of $\frac{d\Phi}{dt}$ for the IMLPG_R particles are estimated from the FEM solution using the interpolation scheme mentioned for Eq. (11.25).

The pressure gradient for computing the velocity within the IMLPG_R domain can be evaluated by using the SFDI (Ma, 2008) from the solution of pressure in the IMLPG_R domain. The problem is that this scheme is based on a local support domain associated with each particle. For a particle near B_1, its local support domain may go outside of the IMLPG_R domain. To evaluate the pressure gradient more accurately, several layers of imaginary particles, called as "Feeding particles" are added to the left of B_1 as shown in Fig. 11.2. The domain associated with the feeding particles is called as D_2 and its upstream and downstream boundaries are B_0 and B_1, respectively. The position of the particles is determined by using the velocity from the solution in the FEM domain. The pressure values at the feeding particles are estimated in the same way as for finding the pressure of the particles on B_1 given in Eq. (11.26), i.e., also from the FEM solution. Based on our experience, at least four layers of feeding particles are required. In our test cases reported later, 10 layers of feeding particles are used to model the steep wave cases.

11.4.1.3   *Treatment of the velocity in the overlapping zone*

After finding the solution at each time step, the free surface in the FEM region should be updated and the positions of the particles for the IMLPG_R method should also be updated. This involves the need of velocity estimation, which can be calculated from the velocity potential in the domain (D_1 and D_2) and from pressure in D_4 (Fig. 11.2(b)). The dilemma exists in the overlapping zone (D_3). Within the overlapping zone, there are nodes for the FEM method and particles for the IMLPG_R method. The velocity field within this zone determined by the solution in the FEM domain may not be exactly the same as that determined by the solution in the IMLPG_R domain. Considering this fact, the particles in this region of the IMLPG_R zone are called "two layer particles" (TLPs). As long as there are no broken waves in the zone, the viscous effect may not play an important role and so both solutions should be close to each other. Nevertheless, use of either one will lead to discontinuity across boundaries B_1 and B_2. Even the quantity of discontinuity is small, the spurious high frequency wave components can be induced as quoted in Sitanggang (2008). To overcome this difficulty, it is proposed to smooth the velocity in the overlapping zone before updating the free surface and the position of particles for the IMLPG_R method. Specifically, the velocity at the TLPs in the overlapping zone (D_3) is modified using the following equation:

$$\vec{u}_p = (1 - \alpha)\vec{u}_{\text{FEM}} + \alpha\,\vec{u}_{\text{IMLPG\_R}}, \tag{11.27}$$

where α is given by

$$\alpha = \begin{cases} 1 & s/L_0 \leq 0 \\ 1 - \sin\left[\dfrac{\pi}{2}\dfrac{s}{L_0}\right], & 0 < s/L_0 < 1 \\ 0 & s/L_0 \geq 1, \end{cases}$$

$$L_0 = x_{\text{FEM}} - x_{\text{IMLPG\_R}}, \quad s = x_{\text{FEM}} - x_p$$

$$x_{\text{IMLPG\_R}} = \min(x_f, x_b)$$

where x_{FEM}, x_f and x_b are depicted in Fig. 11.2. and x_p is the coordinate of the particle under considerations inside the overlapping zone. The velocity $\vec{u}_{\text{FEM}}$ at the particles for the IMLPG_R method is obtained by interpolating the solution of the FEM method using the MLS scheme mentioned already.

11.4.2 Coupling methods in time domain

In the time domain, one can use either an iterative procedure, which is common in fluid-structure interaction problem or an implicit coupling between these two computational domains. In using the first option, the convergence needs to be checked at the end of every time step. The two solvers mentioned earlier work with different time-integration methods, i.e., in the FEM solver an explicit time integration is used (Eq. (11.8)), while in the IMLPG_R fully implicit time integration is employed (Eq. (11.22)). Thus, in the proposed hybrid method, the exchange of data from one computational domain to the other is accomplished through a second-order Runge–Kutta method nested with the predictor–corrector algorithm as shown in Fig. 11.3. This is a kind of implicit coupling.

From Fig. 11.3, one could see that the IMLPG_R solver needs to be invoked twice in each time step. Following Sriram and Ma (2012), the fluid particle positions were unchanged during each time step. This is justified by the fact that the movement of the node is very small in one time step and so change of particle positions caused by the deformation of the domain interface in one step is negligible. Our numerical tests indicate that the solution found in this way is much more stable than those if the particle positions are allowed to vary during the predictor–corrector step. In addition, keeping the fluid particle position unchanged has the following advantages:

(1) Matrix **A** in Eq. (11.24) remains unchanged. The computational time in re-assembling the "**A**" matrix during the correction step is not required.
(2) In our IMLPG_R method, the pressure equations are solved using an iteration solver like GMRES and Gauss–Seidel. Hence, with an unchanged matrix A, the number of iterations required for solving the pressure equations will be minimized.

More discussion on the procedure in Fig. 11.4 is given here. Say at a particular time, "t^n", the simulation from both the computational domains is finished and the value of velocity potential (ϕ), time derivative of potential

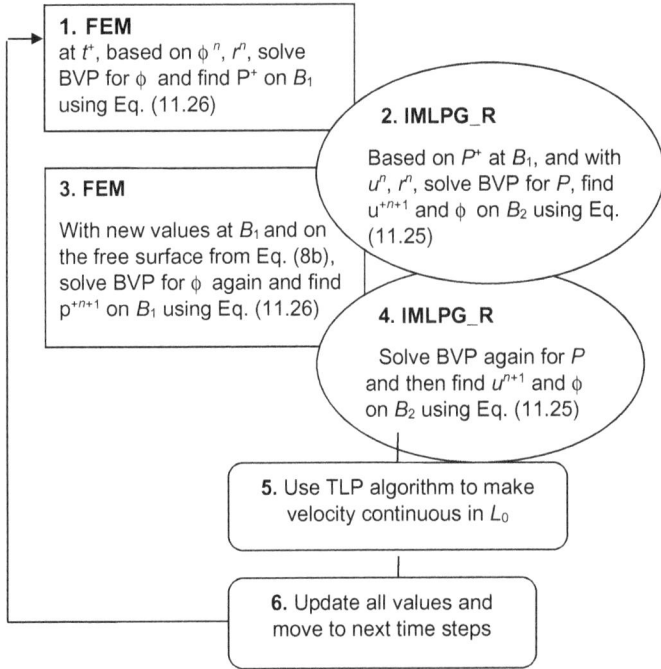

Fig. 11.3   Flowchart of the coupling adopted in time.

$(d\phi/dt)$ velocity, pressure and nodal positions are known at the free surface and boundaries. Then the simulation from "t^n" to "t^{n+1}" follows the subsequent seven stages of calculation, referring to Fig. 11.2(b) and 11.4:

(1) The computation in the FEM domain:

    (a) Solve the BVP for the potential (ϕ^n) using Eq. (11.7) on the D_1, D_2 and D_3 with the known values at F_1, F_2, F_3 and B_2 from the previous time step.

    (b) Solve the BVP to find ($d\phi^{+n}/dt$) on D_1, D_2 and D_3.

    (c) Estimate the intermediate pressure on B_1 (Eq. (11.26)), B_0 and D_2 (Eq. (11.6)) as well as the velocity in D_2 from the FEM solution.

(2) Prediction computation in the IMLPG_R domain:

    (a) Solve the BVP about the pressure in D_3 and D_4, with Dirichlet boundary condition on B_1.

    (b) Find the intermediate velocity in D_3 and D_4 (Eq. (11.19)) with the help of "Feeding particles" in D_2.

   (c) Transfer the data to the nodes on the boundary B_2 for the FEM. Evaluate dynamic pressure using the SFDI scheme.

   (d) Calculate the value of the velocity potential on B_2 using Eq. (11.25).

(3) Intermediate updating in the FEM domain and re-computation:

   (a) Evaluate the intermediate coordinates and potential using Eq. (11.8b) on F_1, F_2 and F_3 and then solve the BVP for the new potential ϕ^{n+1}.

   (b) Solve the BVP to find $(d\phi^{(n+1)}/dt)$.

(4) Estimate the pressure on B_0, B_1 and in D_2 as well as the velocity in D_2 from the FEM solution; then transform them to the IMLPG_R particles on B_1 and on the feeding particles.

(5) Correction computation in the IMLPG_R domain:

   (a) Re-solve the BVP about the pressure in D_3 and D_4.

   (b) Find the velocity in D_3 and D_4 with the help of "Feeding particles" on D_2.

   (c) Transfer the data to the nodes on B_2 for the FEM using Eq. (11.25).

(6) Modify the velocity in D_3 using the algorithm (Eq. 11.27).

(7) Full updating in both domains:

   (a) Evaluate $\vec{r}^{n+1}$ and ϕ^{n+1} using Eq. (11.8d) on F_1, F_2 and F_3 for the FEM domain.

   (b) Evaluate $\vec{r}^{n+1}$ using Eq. (11.22) for the IMLPG_R domain (D_3 and D_4).

   (c) Evaluate the potential and positions of nodes on B_2 using Eq. (11.25).

   (d) Go to next time step.

11.5   Results and Discussions

In Chapter 8, all the test cases that are carried out in experiments are for constant water depth. In order to investigate the earlier process one could use the validated numerical model based on FNPT. However, there is a limitation in the model based on FNPT, which is based on inviscid potential flow theory, and it was not able to model the process of run-up. It would be better to study this using the hybrid FNPT–NS model based on FEM and MLPG_R (Sriram et al., 2014). In this chapter, run-up has also been studied by means of analytical solutions (Didenkulova, 2009) and differently transformed long waves were investigated on different slopes, this would be much faster than the hybrid FNPT–NS and will provide a good preliminary estimate.

11.5.1   *Weak coupling FNPT with analytical run-up model*

The sea surface elevation at $x = 225$ m of the solitary pulses with amplitudes 4 cm, 6 cm and 12 cm (see Chapter 8) are used as input waves to study wave run-up on a beach using the analytical model of Didenkulova (2009). Figure 11.4 shows the corresponding inundation in 1:6 slope, which gets expectedly more complicated with an increase in amplitude of the incident wave and strongly depends on the wave transformation during its propagation. It can be seen that the inundation length increases nonlinearly with an increase in the initial wave amplitude. This is even more visible in Fig. 11.5, where the normalized run-up oscillations are displayed. From Fig. 11.5 one can see that the amplification factor (the ratio of the maximum run-up height to the maximum wave height at the toe of the slope) increases with

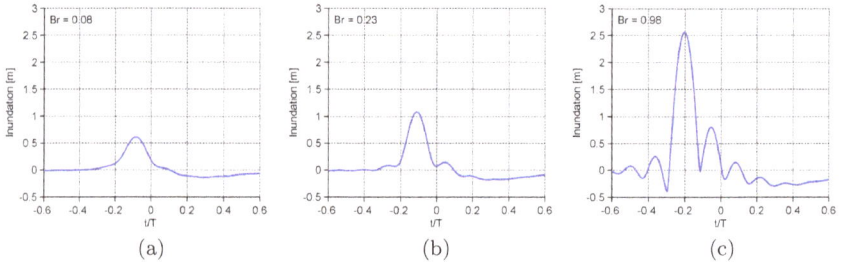

Fig. 11.4   Inundation and run-up speed. (a) $H = 0.04$ m, (b) $H = 0.06$ m and (c) $H = 0.12$ m.

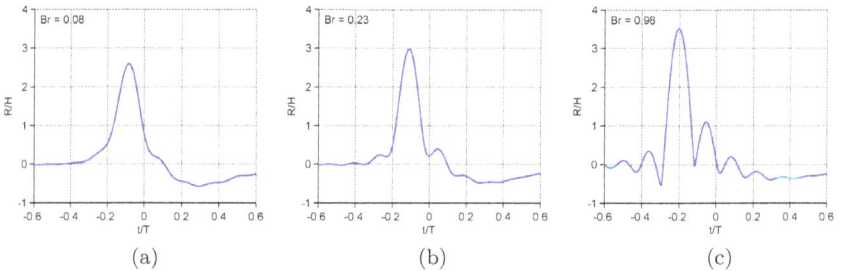

Fig. 11.5   Normalized run-up height in a mild slope of 1:6. (a) $H = 0.04$ m, (b) $H = 0.06$ m and (c) $H = 0.12$ m.

an increase in the amplitude of the initial pulse. So, the wave amplification on the beach for 0.04 m wave is 2.5, for 0.06 m wave is 2.8 and for 0.12 m wave is 3.4.

Parameter Br plotted in Figs. 11.4 and 11.5 is the wave-breaking parameter:

$$Br = \frac{\max\left(\frac{d^2 r}{dt^2}\right)}{g\alpha^2} \leq 1, \tag{11.28}$$

which indicates wave breaking during its run-up. Here r is the run-up oscillations on the beach, g is the gravity acceleration. If $Br < 1$, the wave is not breaking. For $Br = 1$ the wave experiences first breaking just at the beach. This occurs in the back-wash stage and is manifested by the bubble formation when the water is at its maximal offshore position.

11.5.1.1   *Effect of the slope*

Analysis on the effect of the beach slope has been studied hypothetically considering the possibility with different inclinations. For this we studied propagation of the initial pulse in the basin of constant depth (1 m) and distance L (from the wave generator to the beginning of the slope) and then its run-up on the slope α. The propagation was performed using FNPT–FEM up to the slope, which proved to be good for description of long wave dynamics in GWK (as shown in Chapter 8). Run-up height was calculated based on nonlinear shallow water theory as discussed in the previous sections. Both slope α and distance L have been changing in order to keep the total length of the basin constant (see Fig. 11.6).

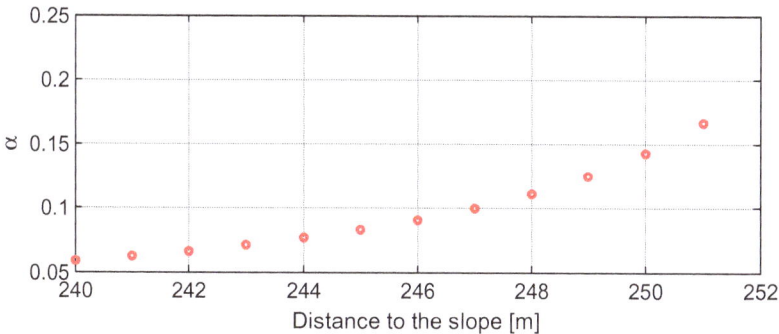

Fig. 11.6   Beach slope α with respect to the distance L.

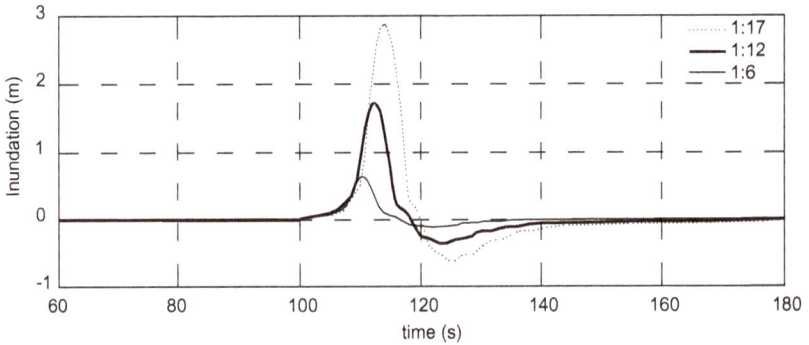

Fig. 11.7   Inundation of 4 cm high elongated solitary pulse for three types of the beach slope: 1:6, 1:12 and 1:17.

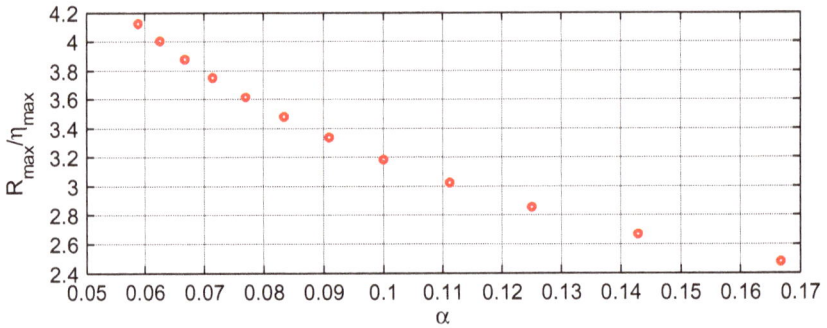

Fig. 11.8   Amplification ratio of 4 cm wave on a beach with slope α.

The result of this analysis is shown in Fig. 11.7 for the initial 0.04 m pulse which climbs three different beach slopes: 1:6 slope (assumed due to the existing GWK beach absorber slope), 1:12 and 1:17. It can be seen that the inundation distance increases substantially with a decrease in the beach slope. For example, the difference in inundation distance on 1:6 and 1:17 slopes is almost 70%.

Figure 11.8 demonstrates that the amplification ratio of the solitary wave on a beach changes with the corresponding change of the beach slope. With this analysis based on analytical solutions of shallow water theory we are limited to the non-breaking wave propagation and run-up, which corresponds to values of $Br \leq 1$. The breaking case will be dealt later using the numerical model.

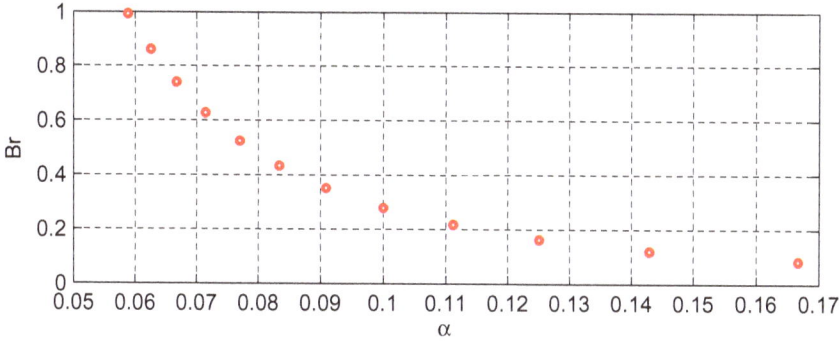

Fig. 11.9 Breaking parameter for the 4 cm wave on the beach with different slopes α.

Behaviour of the breaking parameter Br for 4 cm wave climbing different slopes is shown in Fig. 11.9. It can be seen that even 4 cm wave at the slope 1:17 is already at the point of breaking. The wave of 12 cm will break already at the slope of 1:7 just a little bit milder than the current GWK slope. For further decrease in the beach slope, the wave may form steep front with plunging breaking and bore formation. So, in this way all the three types of waves according to the classification of Shuto (1985) are feasible to be reproduced in the experimental facility.

11.5.2 *Hybrid coupling FNPT with NS*

To study the evolution of long waves as it propagates in the flume, the tests were carried out. A flat bottom was considered during the experimental study. The waves were generated using the piston-type wave maker and the wave elevations were measured at various locations along the flume. Wave probes were placed at various locations along the length of the flume as shown in Fig. 11.10.

The wave elevations obtained from the experiments were compared with the values obtained from the numerical model for verification. It can be observed from Fig. 11.11 that the wave elevations match well in space and time for the test case. The waves tend to steepen as it propagates over the distance of 225 m. The symmetry of the wave has been affected during the propagation due to the dispersion characteristics.

Similar tests were performed for other test cases and the results are presented in the subsequent paragraphs.

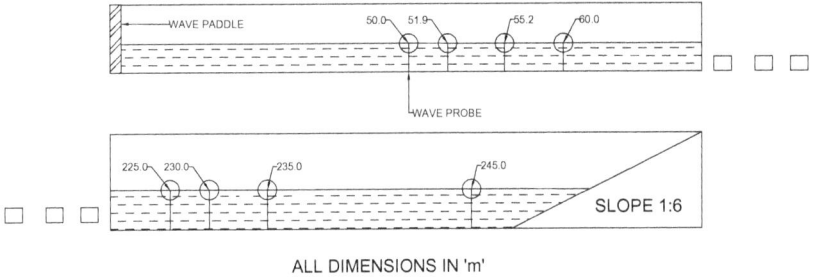

Fig. 11.10   The sketch of the physical model used for the study. Wave probes and locations are marked. Total length is 300 m.

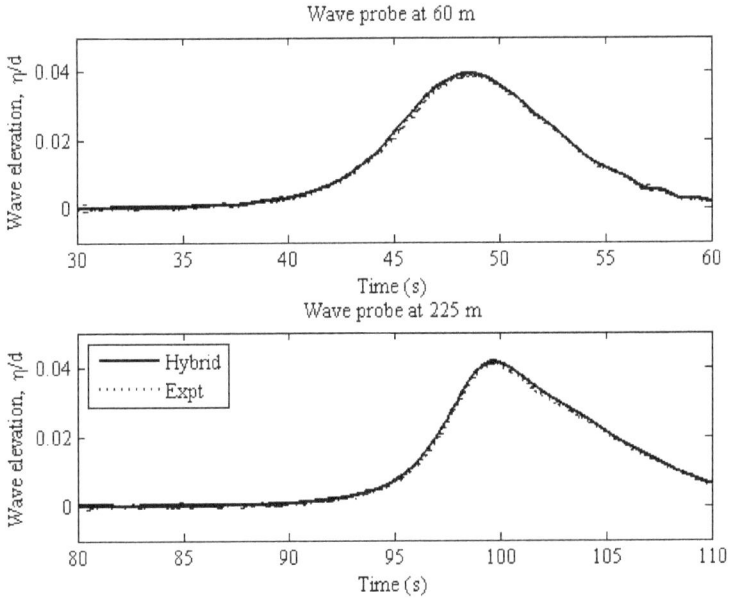

Fig. 11.11   The comparison of wave elevation recorded at different locations along the flume with the hybrid numerical model for an impulse type wave with $H = 0.04$ m.

In Figs. 11.11 and 11.12, the propagation and evolution of the wave has been captured well using the hybrid numerical model. The N-wave propagation is shown in Figs. 11.13 and 11.14, corresponding to small steepness and large steepness. The reason for the deviation may be due to the dispersion characteristics and viscous characteristics, since they have not been properly modelled in the present hybrid model for the steep waves. Further,

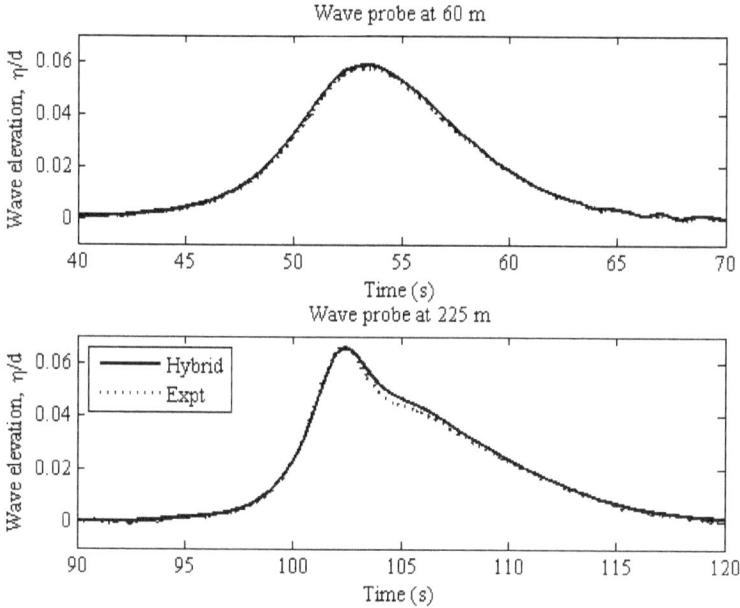

Fig. 11.12   The comparison of wave elevation recorded at different locations along the flume with the hybrid numerical model for an impulse type wave with $H = 0.06$ m.

the number of nodes near to the support domain for steep waves in hybrid model (particularly in the MLPG_R zone) leads to numerical damping. This needs further investigations.

A typical spatial profile of the wave propagation over a slope of 1:40 at various instant of time using the present model is shown in Fig. 11.15. The evolution of the free surface has been depicted in Fig. 11.15.

As mentioned earlier, the empirical formulas developed from the experiments (Synolakis, 1987) form a reliable tool for estimating the run-up for solitary waves in Eq. (11.29). It was observed during the studies presented in Sriram *et al.* (2016) that Eq. (11.30) tend to overpredict the value of run-up when used for certain test cases. Instead of a single parameter empirical formula, a two-parameter model (Eq. 11.30) was suggested.

$$\frac{R_{max}}{H} = 2.8312\sqrt{\cot \alpha}\left(\frac{H}{d}\right)^{1/4}. \tag{11.29}$$

$$\frac{R_{max}}{\eta_{max}} = 2.8312\sqrt{\cot \alpha}\left(\frac{1}{gd}\left(\frac{4\pi d}{\sqrt{3}T}\right)\right)^{1/4}. \tag{11.30}$$

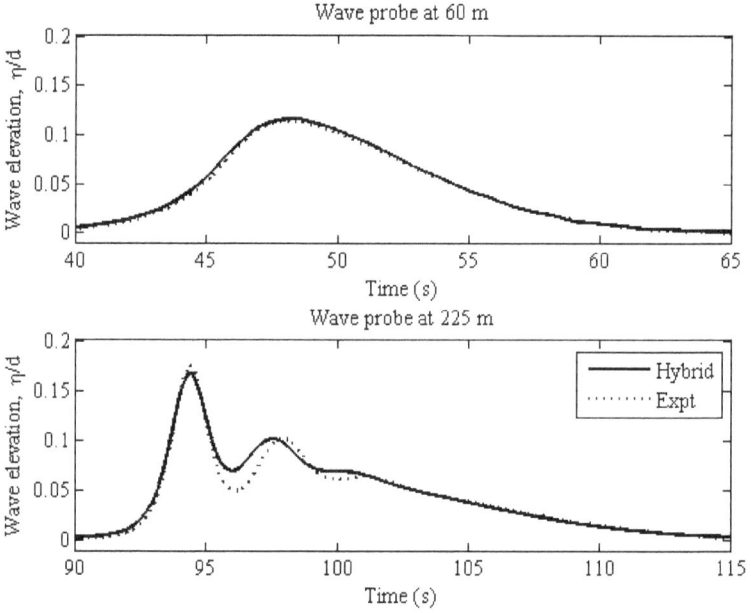

Fig. 11.13   The comparison of wave elevation recorded at different locations along the flume with the hybrid numerical model for an impulse type wave with $H = 0.12$ m.

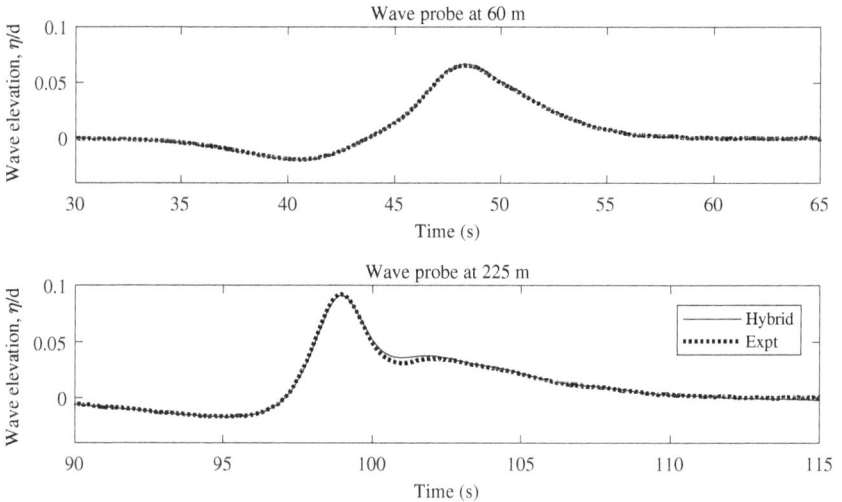

Fig. 11.14   The comparison of wave elevation recorded at different locations along the flume with the hybrid numerical model for an N-wave with $H = 0.09$.

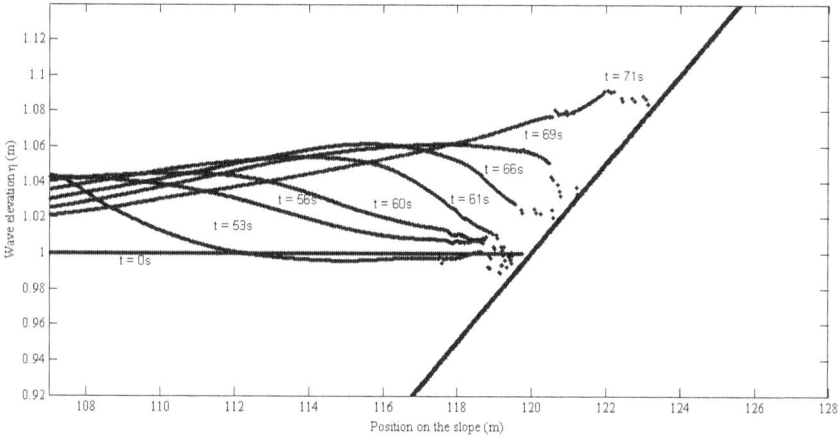

Fig. 11.15   The wave propagation and run-up of case *elongated pulse EP04* over a slope of 1:40.

Table 11.1.   Run-up estimated from the hybrid numerical model for various cases with $H = 0.04\,\mathrm{m}$ and various slopes.

S. No	Case	Slope	$R_{\max}/H$
1	EP04, $T = 30\,\mathrm{s}$	1:10	3.5
2	EP04, $T = 30\,\mathrm{s}$	1:20	2.5
3	EP04, $T = 30\,\mathrm{s}$	1:30	3.25
4	EP04, $T = 30\,\mathrm{s}$	1:40	3.75
5	EP04, $T = 30\,\mathrm{s}$	1:50	4.0
6	EP04, $T = 30\,\mathrm{s}$	1:60	4.175
7	EP04, $T = 100\,\mathrm{s}$	1:30	2.33
8	EP04, $T = 100\,\mathrm{s}$	1:60	3.33
9	EP04, $T = 100\,\mathrm{s}$	1:70	3.95

Tests cases simulated in the hybrid model are presented in Table 11.1 along with the estimated run-up. The run-up over the slope was estimated by tracking the propagation of the free surface along the slope. Even though analytical models are available, they cannot be applied for breaking-wave cases, as discussed previously after 1:17 slope the waves tend to break.

Figure 11.16 shows the run-up measurements from the present hybrid model for the non-breaking and breaking cases. The line representing Eq. (11.29) and $H = 0.04\,\mathrm{m}$ in Fig. 11.16, corresponds to the solitary waves

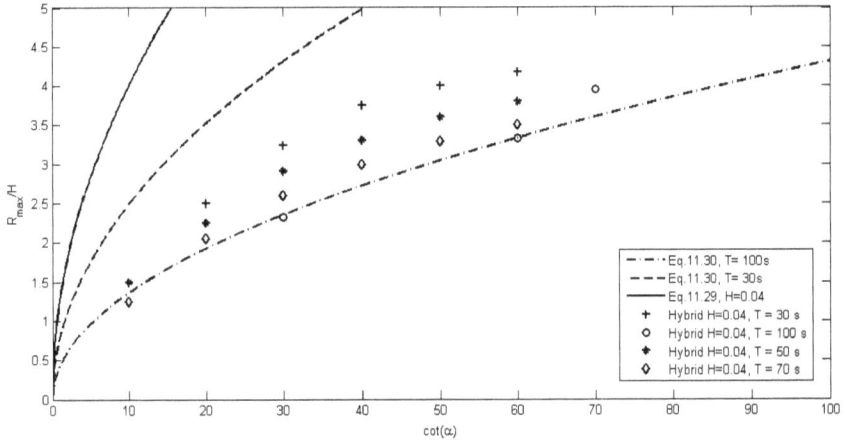

Fig. 11.16 Comparison of run-up estimations from empirical formulas and numerical simulations.

in the steep slope, when the slope is mild there will be the breaking and hence the run-up will be less, which is obtained from the present numerical model (using $H = 0.04$ and $T = 30\,\text{s}$). If one increases the wave period to 100 s, the cases are non-breaking and using our hybrid model, we obtain good agreement. Thus, the advantage of hybrid numerical model is that it can be applied for all sorts of breaking and non-breaking waves.

11.6   Summary

Dynamics of strongly nonlinear waves including its breaking, formation of shock waves and modelling the run-up aspects using hybrid model has been reported. Two types of modelling can be handled. The first model combines the fully nonlinear potential flow theory for modelling propagation and evolutions and for modelling run-up one can use the analytical model from shallow water equations. The limitation of this model is that it will not work for the breaking cases, and in these scenarios one can go for the hybrid model combining the fully nonlinear potential flow theory with Navier–Stokes equations; this will handle both breaking and non-breaking process. Further, the chapter discusses the limitation of the one-parameter solitary wave for such a complex multi-parameter phenomenon like tsunami using the numerical models.

References

Didenkulova, I. (2009). "New trends in the analytical theory of long sea wave runup" in E. Quak and T. Soomere (eds.), *Applied Wave Mathematics: Selected Topics in Solids, Fluids, and Mathematical Methods*, Springer, Berlin, 265–296.

Grilli, S. T. (2008). "On the development and application of hybrid numerical models in nonlinear free surface hydrodynamics." Keynote lecture in the *Proceedings of the 8th International Conference on Hydrodynamics* (Nantes, France, September 2008) (P. Ferrant and X.B. Chen (eds.)), 21–50.

Kassiotis, C., Ferrand, M., Violeau, D., and Rogers, B. D. (2011). "Coupling SPH with a 1-D Boussinesq-type wave model." 6th International SPHERIC Workshop, Hamburg, Germany.

Khayyer, A., and Gotoh, H. (2009). "Modified moving particle semi-implicit methods for the prediction of 2D wave impact pressure." *Coastal Engineering*, 56(4), 419–440.

Kim, C. H., Clement, A. H., and Tanizawa, K. (1999). "Recent research and development of numerical wave tanks — A review." *International Journal of Offshore and Polar Engineering*, 9(4), 241–256.

Koshizuka, S., and Oka, Y. (1996). "Moving-particle semi-implicit method for fragmentation of incompressible fluid." *Nuclear Science and Engineering*, 123, 421–434.

Ma, Q. W. (2005). "Meshless local Petrov — Galerkin method for two-dimensional nonlinear water wave problems." *Journal of Computational Physics*, 205(2), 611–625.

Ma, Q. W., Wu, G. X., and Eatock Taylor, R. (2001). "Finite element simulation of fully non-linear interaction between vertical cylinders and steep waves. Part 1: Methodology and numerical procedure." *International Journal for Numerical Methods in Fluids*, 36(3), 265–285.

Ma, Q. W. (2008). "A new meshless interpolation scheme for MLPG_R method." *Computer Modeling in Engineering and Sciences*, 23(2), 75–89.

Ma, Q. W., and Yan S. (2006). "Quasi ALE finite element method for nonlinear water waves." *Journal of Computational Physics*, 212(1), 52–72.

Ma, Q. W., and Zhou, J. T. (2009). "MLPG_R method for numerical simulation of 2D breaking waves." *CMES*, 43(3), 277–303.

Ma, Q. W., and Yan, S. (2009). "QALE-FEM for numerical modelling of nonlinear interaction between 3D moored floating bodies and steep waves." *International Journal for Numerical Methods in Engineering*, 78, 713–756.

Madsen, P. A., Fuhrman, D. R., and Schäffer, H. A. (2008). "On the solitary wave paradigm for tsunamis." *Journal of Geophysical Research*, 113, C12012.

Manoj Kumar, G., Sriram, V., and Schlurmann, T. (2017). "Propagation and breaking characteristics of solitons and N-wave in fresh water and brine." *Journal of Hydraulic Research*, 55(4), 557–572.

Schäffer, H. A. (1996). "Second-order wavemaker theory for irregular waves." *Ocean Engineering*, 23(1), 47–88.

Schimmels, S., Sriram, V., and Didenkulova, I. (2016). "Tsunami generation in a large scale experimental facility." *Coastal Engineering*, 110, 32–41.

Shuto (1985). "The Nihonkai–Chuubu earthquake tsunami on the north Akita coast." *Coastal Engineering Japan, JSCE*, 28, 255–264.

Sitanggang, K. I. (2008). "Boussinesq-equation and RANS hybrid wave model." Doctoral thesis, Texas A&M University, USA.

Sriram, V., Sannasiraj, S. A., Sundar, V., Schlenkhoff, A., and Schlurmann, T. (2010). "Quantification of phase shift in the simulation of shallow water waves." *International Journal for Numerical Methods in Fluids*, 62(12), 1381–1410.

Sriram, V. (2008). "Finite element simulation of nonlinear free surface waves." Doctoral Thesis, Department of Ocean Engineering, IIT Madras, India.

Sriram, V., and Ma, Q. W. (2012). "Improved MLPG_R method for simulating 2D interactions between violent waves and elastic structures." *Journal of computational Physics*, 231, 7650–7670.

Sriram, V., Didenkulova, I., Sergeeva, A., and Schimmels, S. (2016). "Tsunami evolution and run-up in a large scale experimental facility." *Coastal Engineering*, 111, 1–12.

Sriram, V., Ma, Q. W., and Schlurmann, T. (2014). "A hybrid method for modelling two dimensional non-breaking and breaking water waves." *Journal of Computational Physics*, 272, 429–454.

Sriram, V., Sannasiraj, S. A., and Sundar, V. (2006). "Numerical simulation of 2D nonlinear waves using finite element with cubic spline approximation." *Journal of Fluids and Structures*, 22(5), 663–681.

Sueyoshi, M., Kihara, H., and Kashiwagi, M. (2007). "A hybrid technique using particle and boundary-element methods for wave-body interaction problems." 9th International conference on numerical ship hydrodynamics, Ann Arbor, Michigan.

Synolakis, C. E. (1987). "The runup of solitary waves." *Journal of Fluid Mechanics*, 185, 523–545.

Yan, S. (2006). "Numerical simulation of nonlinear response of moored floating structures to steep waves." Doctoral thesis, School of Engineering and Mathematical Sciences, City University, London.

Zhou, J. T., and Ma, Q. W. (2010). "MLPG Method based on Rankine source solution for modelling 3D Breaking Waves." *Computer Modeling in Engineering & Sciences (CMES)*, 56(2), 179–210.

Zhou, J. T., Ma, Q. W., and Zhang, L. (2009). "Numerical investigation on violent wave impacts on offshore wind energy structures with meshless method." *Proceedings of the Nineteenth International Offshore and Polar Engineering Conference*, Osaka, Japan, 503–509.

Chapter 12

Tsunami Impact Modelling

12.1 Introduction

Designing coastal and onshore structures is turning out to be challenging when we include tsunami-induced forces. The tsunami of March 2011 that hit the north-eastern coast of Japan and the tsunami of December 2004 that occurred in the Indian Ocean have exposed the limitations in the engineering design of coastal and onshore structures. The conventional critical load combination and the factor of safety adopted for each type of loads are not sufficient to withstand the impact of tsunami.

The guidelines for the design of structures exposed to tsunamis have been detailed by Federal Emergency Management Agency's Coastal Construction Manual, FEMA (2005), and the City and County of Honolulu Building Code, CCH (2000). Asakura *et al.* (2000) have carried out laboratory tests on the forces on structures exposed to tsunami, the results of which were adopted by Okada *et al.* (2005) to establish design guidelines. Yeh (2007), through a critical review of the existing literature, proposed rational methodologies for the estimation of tsunami-induced loadings. In general, tsunami loads are considered in the following forms: hydrostatic, buoyant, hydrodynamic, surge, debris impact and wave-breaking forces. Among these, the impact force due to wave breaking has been defined inadequately. The plunging type breaking occurs in a finite depth of water and, once the wave crest plunges, it becomes a broken wave in a shallower depth. The broken tsunami wave moves like a *bore* because of its long wavelength. The movement of tsunami on a dry land is called *surge*. In this chapter, the impact of tsunami on coastal and onshore structures is considered as *the* impact of a bore front, considering the situation of the wave breaking occurring further offshore.

The bore impact on a wall was numerically predicted by Mohapatra *et al.* (2000) by solving two-dimensional flow equations. The predictions were compared with the experimental data given by Ramsden (1993). In this chapter, the bore impact force on a vertical structure is explored using the pressure impulse theory proposed by Cooker and Peregrine (1995). The pressure impulse theory is useful for predicting high pressures that result from the sudden velocity changes at the impact.

The pressure impulse theory estimates the impulse, which is an integrated quantity. For the design of structures, the prediction of maximum impact pressure, consequently the force, is formidable, since such impact takes place in a very short duration, of the order of milli-seconds. Hence, the impulse, that is, the integrated energy form, provides a convenient design parameter to estimate the tsunami wave impact.

The following sections present the pressure impulse theory in the context of bore impact and the subsequent validation of the theory for a tsunami impact on a nearshore elevated structure.

12.2   Pressure Impulse Theory

12.2.1   *General*

The pressure impulse (P) is defined in terms of instantaneous impact pressure, $p(x, t)$ within a short duration, $t_a < t < t_b$.

$$P = \int_{t_a}^{t_b} p(x, t)dt. \tag{12.1}$$

In a two-dimensional flow field, with the assumption of inviscid and incompressible fluids, Euler's equation of motion can be used to model the wave impact phenomenon. The nonlinear convective terms in the equation of motion have been neglected, under the assumption that the large local pressures last only for a shorter duration. Therefore, Euler's equation can be written as follows:

$$\frac{\partial u}{\partial t} = -\frac{1}{\rho}\nabla p, \tag{12.2}$$

where, ρ is the density of water and u is the flow velocity at impact. Integrating Eq. (12.2) over the impact duration $(t_a \leq t \leq t_b)$ and taking the divergence leads to the following Laplace equation:

$$\nabla^2 P = 0. \tag{12.3}$$

Figure 12.1 shows a two-dimensional idealized model of the wave impact on a vertical wall. The fluid domain (Ω) has been idealized with a horizontal free surface (Γ_F) at the top of the bore $(z = 0)$ and a horizontal bottom surface $(z = -h)$. The rigid vertical wall forms the right boundary (Γ_W) at $x = 0$. The total flow depth (h) is taken to be equal to the bore height (d_1).

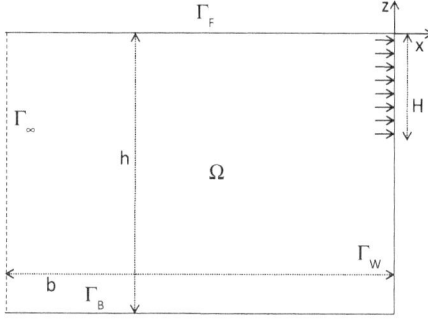

Fig. 12.1   Idealized fluid domain.

The impact on the wall occurs over a depth of $(d_1 - d_2)$, where d_2 is the flow depth in front of the bore movement.

12.2.2   Boundary conditions

The boundary conditions imposed on the fluid domain are defined here.
On the free surface, Γ_F,

$$P = 0. \tag{12.4}$$

The impulse on the far field (left boundary, Γ_∞) vanishes:

$$P = 0. \tag{12.5}$$

Due to the absence of velocity flux along the bottom boundary, the pressure impulse does not dissipate through the bottom boundary, $\Gamma_B(z = -h)$. Hence,

$$\frac{\partial P}{\partial z} = 0. \tag{12.6}$$

On the rigid vertical wall, the variation in normal flow velocity leads to pressure impulse flux on the impact region. Below this region, where the fluid is initially in contact with the wall before and after the impact $(z < d_2)$, the pressure impulse flux vanishes. This assumption can be justified by considering the flow field of a turbulent bore:

$$\frac{\partial P}{\partial n} = \begin{cases} -\rho U_b & -(d_1 - d_2) \leq z \leq 0 \\ 0 & -d_1 < z < -d_2 \end{cases}. \tag{12.7}$$

12.2.3   Numerical procedure

The pressure impulse is computed during the flow impact. The impulse is taken as field variables in the entire domain and the governing Laplace equation (Eq. (12.3)) is solved using the finite element method. The governing equation along with the four boundary conditions may be expressed in the weighted residual form to obtain the finite element system of equations as follows:

$$\sum_e \int_{\Omega^e} \nabla N_i \nabla N_j d\Omega^e \{P_j\} = \sum_e - \int_{\Gamma_w} N_i \frac{\partial P}{\partial n} d\Gamma^e. \qquad (12.8)$$

In the finite element approach, the domain is discretized into sub-domains (Ω^e) represented by discrete elements, which exchange information through the joining nodes. The pressure impulse at any point inside the sub-domain can be represented in the form of a prescribed shape function $N(x, y)$ and is presented as follows:

$$P = \sum_{j=1}^{n_e} P_j N_j(x, y), \qquad (12.9)$$

where P_j denotes the nodal variables and n_e is the number of element degrees of freedom. The shape functions are explicitly written as the functions of quadratic isoparametric coordinates. The summation implies assembly of the element property matrices over the entire domain, Ω (Zienkiewicz and Taylor, 1989) and the resulting simultaneous equations are solved for the pressure impulse at every node, P_i, $i = 1, 2, \ldots, M_n$, taking advantage of the symmetric banded nature of the matrix.

The domain length (b) is a crucial parameter for computation. This influences the accuracy of pressure impulse to a great extent. Thus, determining the length of the domain is an important task for proper computation of pressure impulse as well as for reducing computational complexity. The effect of the domain length on the pressure impulse has been discussed by Cooker and Peregrine (1995) and they recommended the length of the domain (b) be equal to or greater than the height of the domain (h) for proper computation of pressure impulse. In this study, the length of the domain is set equal to the total fluid depth, i.e., $b = h$.

12.3   Results and Discussion

12.3.1   *Impulse on a vertical wall*

To validate our finite element model with the analytical solution of Cooker and Peregrine (1995), we first simulate the pressure impulse on a vertical wall induced by a plunging wave. The fluid impact depth (H) is taken as half the fluid depth. Various domain lengths (b/h) are considered in the range from 0.1 to 100. Pressure impulse is normalized with the fluid velocity at impact (U_b), the depth of fluid impact (H) and the fluid density (ρ). The vertical elevation (z) is normalized with the total fluid depth (h). Figure 12.2 demonstrates that the domain width of the order of fluid depth $(b/h = 1.0)$ is sufficient for the estimation of pressure impulse: the deviation at the peak impulse from the case of $b/h = 100$ is XX%. It is emphasized

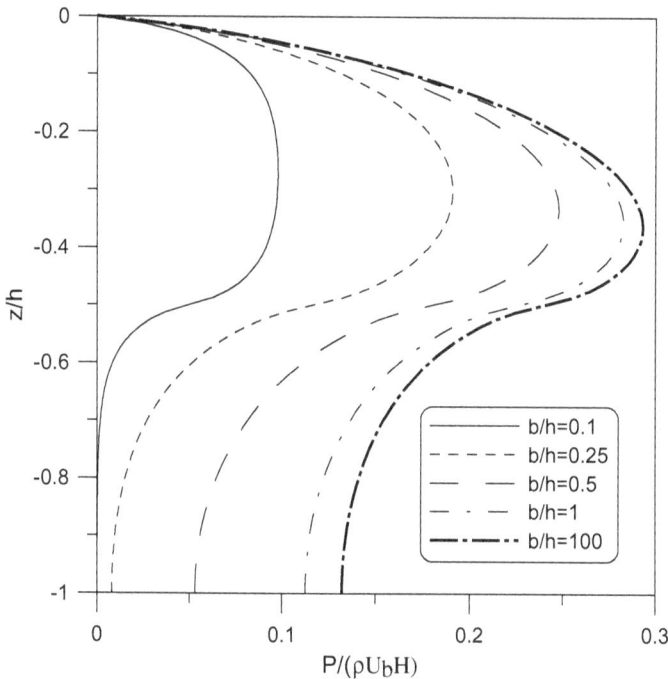

Fig. 12.2   Pressure impulse distribution along the depth of a vertical wall for various breadths of the fluid domain. Height of impact is half the depth of water.

that our numerical prediction matches exactly with the analytical solutions of Cooker and Peregrine (1995).

12.3.2 *Jet impingement on a vertical surface*

The jet impact pressure on an overhanging vertical wall is computed. The height of the elevated wall (d) is taken as being equal to the depth of the fluid domain (h) by assuming free surface condition on the top and bottom of the fluid domain. Figure 12.3 presents the pressure impulse distribution on the wall. A maximum normalized pressure impulse is computed to be 0.36 at the mid-point of the wall. Note that the jet impact is higher than the breaking wave impact on the half depth of the wall presented in the last section. This is simply because the impact velocity U_b is uniformly applied to the jet impingement case, whereas U_b for the plunging breaker case is the maximum value; evidently, the momentum of the latter should be smaller than that of the former.

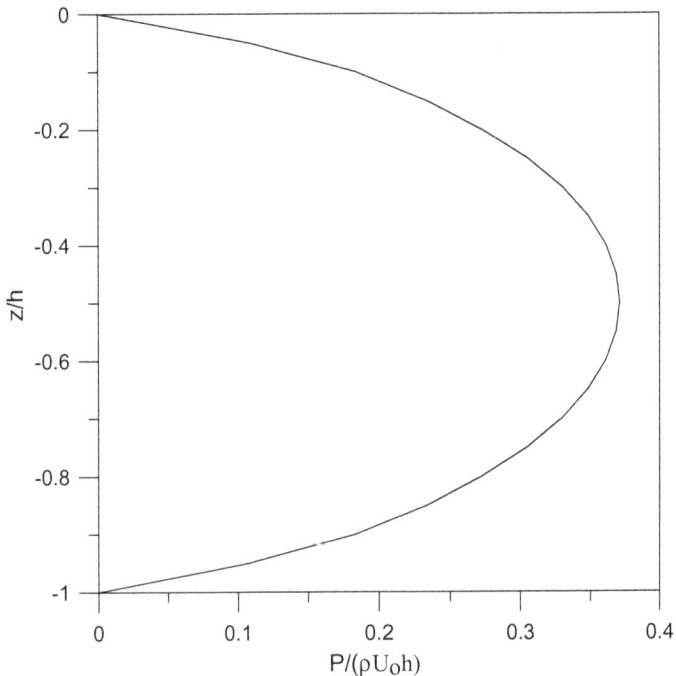

Fig. 12.3   Jet impact pressure on a vertical face of an overhanging wall.

12.3.3 *Tsunami impact on an elevated building*

An attempt is made to compare the numerical prediction with the experimental measurements of impact pressure on an elevated nearshore building. The experiments were carried out in a wave flume of length 104 m, width 3.7 m and depth 4.6 m at the Oregon State University, USA. A piston type wave maker with a maximum stroke of 4 m was used. The experiments were performed by generating a solitary wave of height 0.5 m at a water depth of 2.42 m. The generated wave climbed up the plane beach of 1:12 slope upto the horizontal plateau of the water depth of 0.06 m. The model building was placed on the horizontal section at 3 m from the top of the sloping beach. Note that the incoming wave was broken offshore, forming a turbulent bore before reaching the model. The breadth of the model building along the wave direction is 1.07 m. The bottom clearance of the model building is 0.3 m from the horizontal bed. A pressure transducer was placed on the front face and the model was equipped with a set of force transducers to measure the horizontal force as well as the moment. Figure 12.4 shows the wave profile that was measured with the non-intrusive ultrasonic wave gauge located 3.02 m upstream from the model building. The sampling rate of the pressure and force transducers was 1000 Hz, whereas the wave data were taken at the rate of 50 Hz.

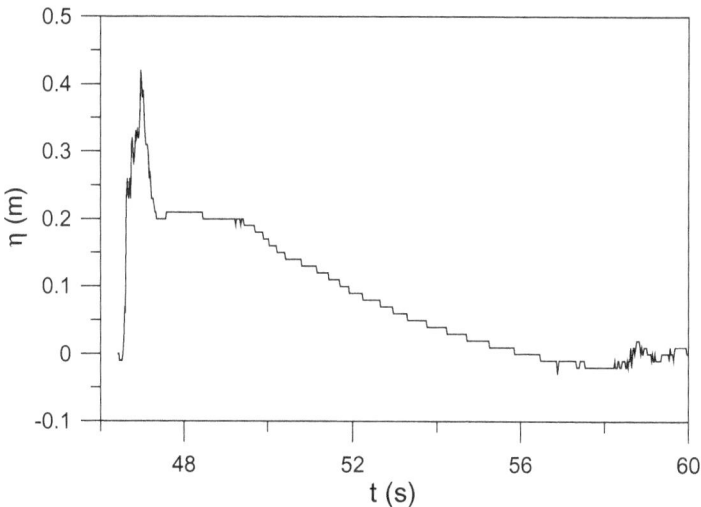

Fig. 12.4   Wave elevation at 3.02 m in front of the building.

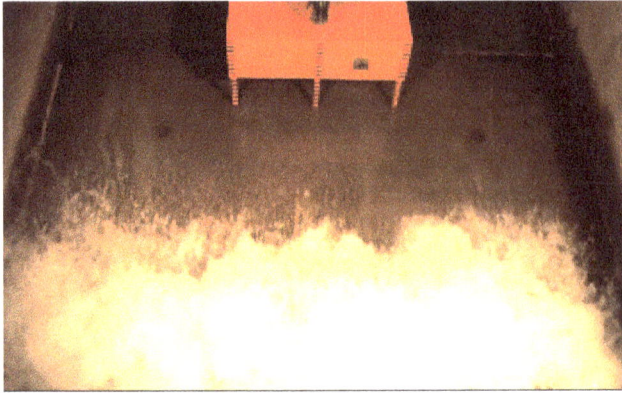

(a)

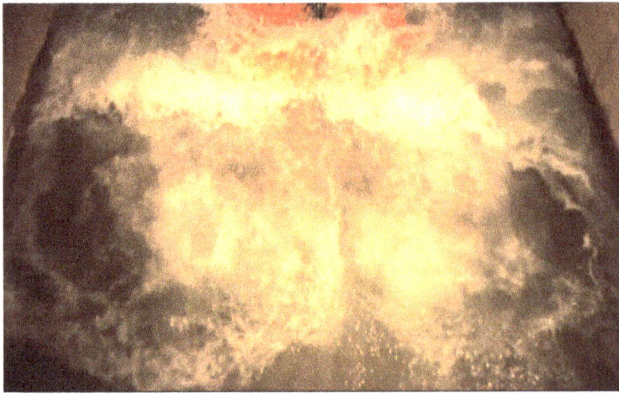

(b)

Fig. 12.5 Bore-type impact on an elevated building. (a) An elevated building at
an initial water depth of 0.06 m. (b) Bore impact on the elevated building.

Figures 12.5(a) and 12.5(b) show the bore front approaching the building and the subsequent impact on the building. The photos were extracted
from the high-speed video data with a frame rate of 75 Hz. The measured
pressure and horizontal force time histories are presented in Fig. 12.6. The
impact pressure can be observed with a rise time of 0.02 s. The pressure
rise–fall time is 0.146 s. The maximum impact occurred at $t = 55.776$ s at
the intensity of 1.489 kN/m^2. The corresponding maximum horizontal force
is 0.744 kN. The data in Fig. 12.6 indicate that the peak response of the
pressure was delayed by 0.2 s. This discrepancy was caused by the location

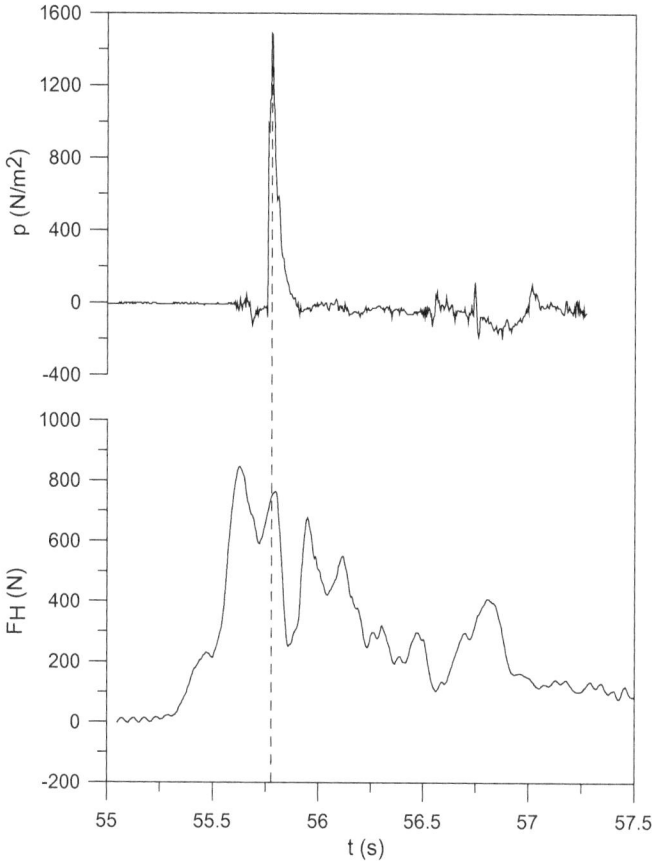

Fig. 12.6   Measured impact pressure on the front face and horizontal force on an elevated building.

of the pressure sensor. The real impact never takes place in the formation of the jet impingement (see Section 12.3.2).

The pressure impulse theory is applied to this problem. The domain used for the finite element computation is projected in Fig. 12.7. The upstream bore height (d_1) is ~0.4 m (estimated from Fig. 12.4) and the initial water depth at the foot of the building is 0.06 m. During the fluid impact on the elevated part of the building (i.e., 0.3 m above the bed), the downstream flow depth was observed to increase and nearly inundate the bottom clearance of the building. This is due to the front face of the bore having a mild slope when it advances into the relatively shallow quiescent water. The

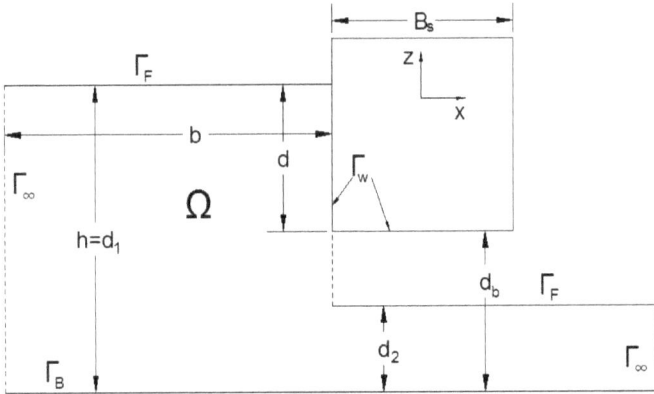

Fig. 12.7   Elevated building model domain.

downstream water depth (d_2) during the bore front impact on the building was assumed to be 0.32 m. The flow velocity at the impingement was found to be 4.03 m/s from the frame-by-frame analysis of the high-speed video footage.

The normalized pressure impulse distribution on the vertical front face of the building is shown in Fig. 12.8. Although the pressure impulse was measured at only one location, it was in good agreement with the computed pressure impulse distribution. The maximum pressure impulse occurred at the bottom of the elevated building.

Table 12.1 presents the horizontal force impulse computed based on the pressure rise–fall time, and the numerical result is compared with the experimental measurement. It can be noted from the force measurement that unlike pressure time history, force time history does not exhibit the formation of the sharp impact.

12.3.4   *Effect of bottom clearance of the building*

A parametric study is performed to estimate the pressure impulse for different bottom clearances of the building. The dimensions and the flow characteristics are assumed to be similar to those mentioned in Section 12.3.3, except that the bottom clearance has been varied from 0 (simulating a vertical wall) to 0.375 m. It is evident that the largest impulse pressure occurs when there is no bottom clearance. A provision of bottom clearance reduces the maximum impact pressure on the building. Figure 12.9 presents a typical variation of pressure impulse on an elevated

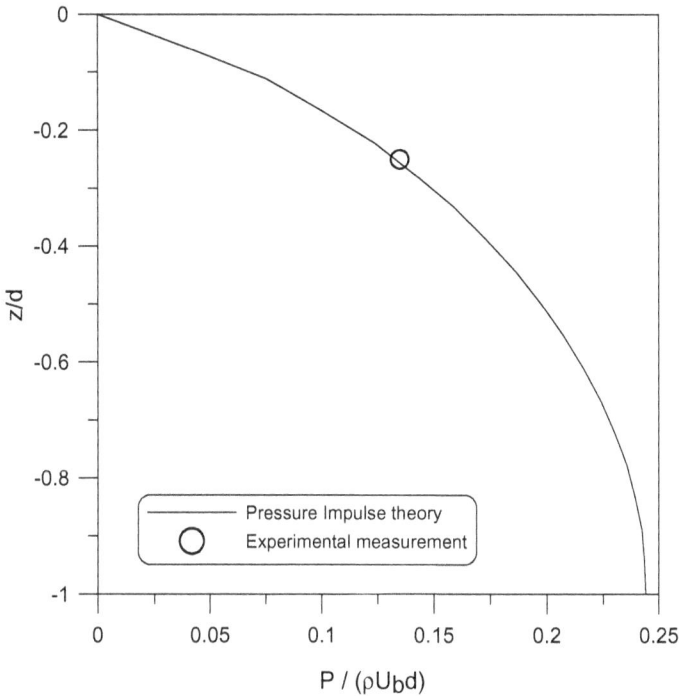

Fig. 12.8   Pressure impulse along the vertical face of the elevated building.

Table 12.1.   Horizontal force impulse.

Horizontal force, $F_I/(\rho U_b T^2 B)$	
Experiment	0.315
Numerical	0.176

building. Here, d_o is the bottom clearance of the building, d_1 is the tsunami front height reaching the building and d is the height of the building from its bottom elevated position up to the maximum tsunami run-up level. $z/d = 0$ indicates the free surface and $z/d = -1$ indicates the bottom wall level. Hence, for $d_o = 0$, i.e., for a building located at the natural ground level, the maximum pressure is induced. The effect of reduction in pressure impulse is significant even when the bottom clearance is small: for the case of $d_b/d_1 = 0.05$, there is a 24% reduction of the maximum pressure impulse from the case of no clearance. A significant decrease in pressure impulse is observed even for a small bottom clearance: for the case of $d_b/d_1 = 0.05$, a

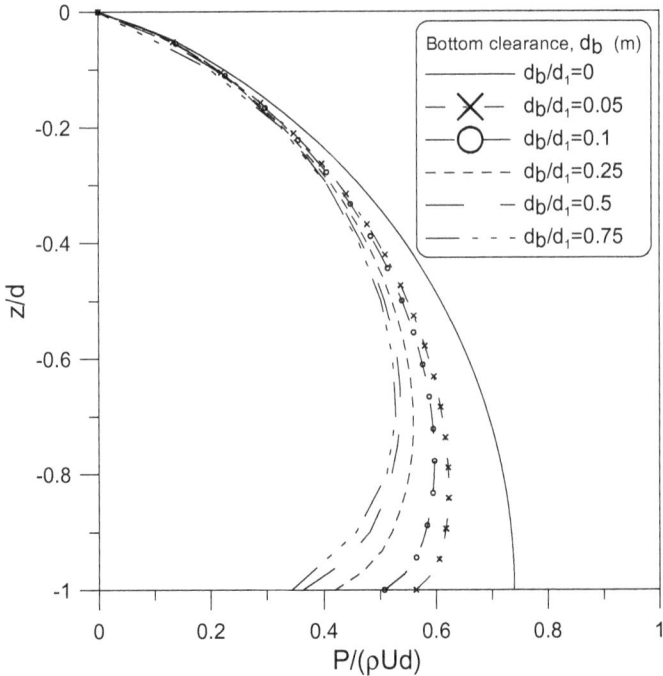

Fig. 12.9   Variation of pressure impulse on the vertical face of the building with different bottom clearances.

reduction of the maximum pressure impulse from the case of no clearance to an extent of about 70% is noted. This estimate has been made using pressure impulse theory and validated by experimental measurements.

12.4   Summary

The tsunami-induced forces on nearshore structures were modelled as the turbulent bore front impact. The pressure impulse theory was adopted and finite element method was used to solve for the pressure impulse on a vertical wall of an elevated building. The comparison between a typically measured long-wave impact on the building and the numerical prediction is found to be in good agreement. The bottom clearance provision for the buildings reduces the pressure impulse to an extent of 20–30%. Furthermore, additional bottom clearance does not substantially reduce the pressure impulse.

References

Asakura, R., Iwase, K., Ikeya, T., Takao, M., Fujii, N., and Omori, M. (2000). "An experimental study on wave force acting on on-shore structures due to overflowing tsunamis." *Proceedings of Coastal Engineering, Japan Society of Civil Engineering*, 47, 911–915.

City and County of Honolulu Building Code (CCH) (2000). "Department of Planning and Permitting of Honolulu Hawaii." Chapter 16, Article 11, Honolulu, Hawaii.

Cooker, M. J., and Peregrine, D. H. (1995). "Pressure-impulse theory for liquid impact problems." *Journal of Fluid Mechanics*, 297, 193–214.

FEMA Coastal Construction Manual (2005). "FEMA 55 Report." Edition 3, Federal Emergency Management Agency, Washington D.C.

Mohapatra, P. K., Bhallamudi, S. M., and Eswaran, V. (2000). "Numerical simulation of impact of bores against inclined walls." *Journal of Hydraulic Engineering*, 126(12), 942–945.

Okada, T., Sugano, T., Ishikawa, T., Ohgi, T., Takai, S., and Kamabe, C. (2005). "Structure design method of buildings for tsunami resistance (proposed)." English translation of The Building Letter of 2004.

Ramsden, J. D. (1993). "Tsunamis: Forces on a vertical wall caused by long waves, bores, and surges on a dry bed." Report No. KH-R-54, W.M. Keck Laboratory, California Institute of Technology, Pasadena, California, 251.

Yeh, H. (2007). "Design tsunami forces for onshore structures." *Journal of Disaster Research*, 2(6), 531–536.

Zienkiewicz, O. C., and Taylor, R. L. (1989). *The Finite Element Method*, Vol. 1, McGraw Hill Book Company, London.

Index